Days of Glory, Days of Gloom

The Rise and Fall of the
Nuclear Power Industry

A. Victor Morisi

PublishAmerica
Baltimore

© 2006 by A. Victor Morisi.
All rights reserved. No part of this book may be reproduced, stored in a retrieval system or transmitted in any form or by any means without the prior written permission of the publishers, except by a reviewer who may quote brief passages in a review to be printed in a newspaper, magazine or journal.

First printing

ISBN: 1-4241-3488-9
PUBLISHED BY PUBLISHAMERICA, LLLP
www.publishamerica.com
Baltimore

Printed in the United States of America

Dedication

This book is dedicated to the countless men and women who pioneered the peaceful use of the atom by working in the nuclear powered electric utility field in the 1960's, 70's and 80's. They were truly another breed of astronauts, they were nuclearnauts, venturing into a new, untried universe, often flying by the seats of their pants to achieve success, always coupled with public safety. They often sacrificed quality time with family and friends to assure safe, reliable power 24/7/365. The author was fortunate enough to be one of those pioneering venturers and this book attempts to give them the thanks and recognition they deserve by writing about their travails and successes. The nuclear power industry, especially in America, was to be the savior of our natural resources until another kind of power source was developed. However, Three Mile Island and Chernobyl derailed those hopes and the nuclear power industry, thirty plus years later, has not recovered its rightful place in our electric power mix. There is movement afoot presently to re-establish the construction of new nuclear power plants. The author also takes the opportunity to define the causes of human burnout of those who toiled to make this power source a viable entity.

This book is also dedicated to my late son, Michael Vittorio Morisi, a brilliant lawyer and a compassionate human being whose life was shortened after a valiant eight-year battle with pancreatic cancer. He was not only my son but also my best pal.
(1957-2000)

Table of Contents

Foreword   9

Chapter 1   15
The Fossil Fuel Mentality at the Top of the House of Most Public Utilities

Chapter 2   19
Nuclear vs. Fossil—The In-House Manpower War

Chapter 3   23
How Do I Get into This Fortress Anyway?

Chapter 4   27
Are You Kidding, Fitness for Duty(FFD)?

Chapter 5   29
Redundancy, Redundancy, Redundancy

Chapter 6   33
Your '65 Ford Now Needs a Catalytic Converter!

Chapter 7   36
Forty Years Old and Just Like New!

Chapter 8   39
One Changes the Light Bulb, One Holds the Ladder and Sixteen Document the Job!

Chapter 9   42
So You Just Made the "Bad Guy List" Huh!

Chapter 10   45
Rising Standards of Excellence or "You Better Row Harder"

Chapter 11   48
Too Many Masters and Not Enough Slaves

Chapter 12   52
Mess with the Bull and You Get the Horns

Chapter 13   57
Documentation Up the Ying Yang!

Chapter 14   60
Configuration Control: Easy to Say but Tough to Implement

Chapter 15   63
The NRC Turns the Screws Up a Notch

Chapter 16   67
The Long Term Plan (LTP), Sanity at Last

Chapter 17   71
The Nitty Gritty of Being on the Bad Guy List

Chapter 18   75
The High Level Radioactive Waste Dilemma in the USA

Chapter 19   78
The Radwaste Concentrator Fiasco

Chapter 20   81
Main Steam Line Isolation Valve Inflatable Test Plugs, Yea or Nay?

Chapter 21	85
Material False Statements or "You Better Tell It Like It Is"	
Chapter 22	88
Do as I Say, Not as I Do—the NRC Credo	
Chapter 23	90
A Month in the Life of a Nuclear Power Plant Manager	
Chapter 24	95
The 80% Comfort Zone, Get In and Get Out	
Chapter 25	97
A Layman's Description of the Three Mile Island Accident (TMI)	
Chapter 26	101
Chernobyl—the USSR's Experiment with Disaster	
Chapter 27	104
So What Protects Us from Radiation Exposure from a Nuclear Power Plant?	
Chapter 28	106
The Browns Ferry Cable Spreading Room Fire and Its Aftermath	
Chapter 29	109
Nuclear Power Plant Operator Training—I Thought You'd Like to Know!	
Chapter 30	112
The Reactor Recirculation Piping and the Drywell Czar	
Chapter 31	118
The Drywell Czar Goes on the Road	

Chapter 32   123
What Happens If Your Reactor Vessel Springs a Leak?

Chapter 33   127
The Nuclear Power Game, the Author and Burnout

Chapter 34   133
Nuclear Power in the Generation Mix in the Future

Foreword

After World War II the U.S. Government enacted the law Title 10 of the Code of Federal Regulations, Part 50, regulating the peaceful use of the atom and established the Atomic Energy Commission (AEC) to license and regulate commercial nuclear reactors for electric power generation. General Electric developed the once-through Boiling Water Reactor (BWR), Westinghouse developed the two loop Pressurized Water Reactor (PWR) as well as did Combustion Engineering and Babcock and Wilcox. These industrial giants were in the forefront of commercial nuclear power reactors for power generation.

Just as with all federal programs, the original scope of regulations grew by leaps and bounds such that the paper work associated with commercial nuclear power required up to ten or more times as many people as originally were aboard for initial commercial operation. After Three Mile Island's mishap, the Nuclear Regulatory Commission (NRC), the successor to the AEC, didn't even bother to go to Congress for revisions to the federal laws, but instead issued Regulatory Guides (which were not law) and made the public utilities comply, often times with time-frames which were unattainable by everyday engineering standards. This very fact, along with Information Notices and Bulletins and Branch Technical Positions, caused manpower increases in the nuclear indusrty which in turn added operating costs and cents per kilowatt of power increases which were passed on to the consumer and the stockholders.

The original commercial "nukes" consisted primarily of Shippingsport, Pennsylvania, Humbolt Bay in California, Big Rock

Point in Michigan and Yankee Rowe in Massachusetts. There were other small nukes in operation in Idaho and Wisconsin but all were small power producers and all are shut down presently.

For the people who support nuclear power generation, this book will reinforce their beliefs in this form of power. For those who are neutral, the contents will expand their base of knowledge of what takes place in these bastions of secrecy. And for those who are anti-nukes, they will find or extrapolate sufficient facts to further their cause for total shutdown of all commercial power reactors in the U.S.A. This book is not written to advance any one cause but is directed toward factual information such that the human burnout factor comes into clear perspective when one compiles the "need" required to satisfy multiple regulatory bodies whose requirements are constantly changing, and are diverse and expanding as well as to maintain the confidence of the general public.

The Federal Government either operates or hires contractors to operate their various nuclear facilities, mostly for the generation of strategic nuclear weapons. The principles of operation are similar to commercial nuclear power reactors, the requirements of protection for employees, neighbours, the water, the air, the earth, etc. are all the same, as mandated by federal law. However, we have all read about and seen on *60 Minutes*, *20/20*, etc. that the facilities operated by or for the federal government are shut down in Colorado, Ohio, Tennessee, and Hanford, Washington. Despite the fact that the government regulates commercial nuclear power, they allow their facilities to circumvent and in some cases, disregard the law relative to the impositions placed upon commercial reactors. If the federal government and/or its contractors had to maintain, operate, record, comply, improve and report their operations with the same degree of compliance as they impose upon commercial nukes, they, in the authors purview, would have all been shut down many years ago and probably not provided sufficient justification for restart. Additionally, many executives could have received fines and possibly even jailed for their non-compliance with the laws. However, because they hid behind the skirts of national defense and national security, they continued to pollute our very country they were pledged to protect. The message is clear—we write the laws, we license, we regulate and therefore "do as I say, not as I do." This is the credo of the federal government in all aspects of its performance and it emanates from congress and the other branches of our government.

The author does not intend to demean the necessity for strict adherence to compliance with the law for the benefit and safety of the populace but merely wishes to highlight the struggle mandated by the federal government to maintain this defense-in-depth concept. The author would only request that the federal government march to the beat of the law when it acts as the operator as well as the regulator. It would therefore understand the discomfort it feels when it has to walk in the shoes of commercial nuclear power personnel.

The source of the contents of this book are gleaned from the authors twenty-seven years experience in commercial nuclear power operation. This experience spanned the gamut of licensing, ground-breaking, design review, construction, testing, operation, maintenance, refuelling and modifications as well as dissatisfaction of some or all of the above by the regulator resulting in shutdown, restructuring/reorganization and restart as well as dealing with a skeptical general public during this entire saga.

Some technical aspects may not be portrayed to the letter. Time and wear have taken their toll on this burned-out author.

Examples of the federal government not adhering to its own regulations can be found by searching the Internet. There was one instance where federal officials, appearing before the House Interior Committee, revealed that radioactive waste was shipped to a dozen or so commercial incinerators not licensed to handle the material. Over a period of several years literally hundreds of tons of hazardous waste were shipped to incinerator sites not on the Department of Energy's list of approved incinerators. Most of the hazardous materials were generated at the Savannah River Nuclear Facility.

Private contractors bid for and receive approval to operate such nuclear facilities for the federal government. Incidentally one hearing before the House Interior Committee revealed that all shipments to commercial sites which contained uranium deleted all references to the material for national security reasons. There was one incident that went on to state that it doubted that the burning of low-level waste was a threat to anybody's health.

Let's for a moment dwell on that last statement. One could postulate that a commercial incinerator, not licensed to burn nuclear waste, might not have a sophisticated system for monitoring radiation exposure to the surroundings near the incinerator. Then how could one state that there was doubt anyone was harmed by these actions?

Compare that to the requirements imposed upon a commercial nuclear power generating station by the federal government. First, a nuclear power plant generates gaseous, liquid and solid low level radioactive waste. The gaseous waste is sent through a long time delay system were it decays and is then released with dilution of air. The utility is required to monitor the time, quantity and radioactive level of all gaseous releases. This type of release will be referenced to again shortly. The liquid releases involve dilution of the liquid to acceptably low levels as mandated by federal law. The liquids are released, highly diluted to the surrounding waters, in this utilities instance that is the ocean. Solid waste is compacted, packaged and shipped to Barnwell, South Carolina, for burial. That is not the end of this saga. The utility is required by law to collect air and liquid samples to assess the amount, if any, of radiation found in the plants surroundings. This collection includes air samples from collection station filters and food stuffs from local farms. Such include cranberries, blueberries, farm grown crops, shellfish, food fish and kelp to name a few. There are probably several other sources. These collected products are tested for radioactive levels and compared to background levels and previous findings. This is all well-and-good but there is more.

This monitoring program is based upon the level or radiation found on the collected samples compared to the naturally occurring radiation found in most materials. Elevated findings indicate that some radiation, well within acceptable regulatory guidelines, has occurred. Did I mention that this program, by law, was instituted before the plant was allowed to operate so as to be able to identify if any increases above baseline readings are observed.

What type of sophisticated program do you think an unlicensed incinerator has?

Chapter 1

The Fossil Fuel Mentality at the Top of the House of Most Public Utilities

When the federal government, via the AEC, was formed to license and regulate nuclear power it naturally evolved that the existing public utilities, with the assistance of the nuclear architects and nuclear steam suppliers (G.E., Westinghouse, Combustion Engineering, etc.) applied for construction permits to design, construct, test and operate these commercial reactors. The public utilities foresaw these power plants as cost effective alternatives to the coal, oil, gas fossil fuel generating stations presently providing the bulk of the power in the U.S.A.

Land was purchased away from high density populations as required by the law. The site was prepared, steam supply design selected, the architect/constructor hired and the Preliminary Safety Analysis Report (PSAR) prepared and submitted to the AEC, who after diligent review, issued a construction permit. The chief executives of these public utilities were, for the past 100 years, heads of companies who understood the workings of a fossil fuel utility. One sited the facility near a source of cooling water to condense the steam and remove the residual heat, provided a continuous source of fuel (oil, gas, coal) and obtained the necessary rights-of-ways to transport the generated power to the nearest transmission grid for dispersal and consumption. Then along comes nuclear power and although the licensing requirements were more stringent and they were dealing with a federal agency in addition to a local state public utility agency, they envisioned nuclear power as a boiler with a 2 years supply of fuel already stored in the boiler rather than fuel supplied constantly from an external source. Everything else, from their perspective, was "old hat". One generated steam by a new process but then it was the same, it drove a steam turbine which spun a generator which produced electrical power which was sold to the consuming public in their service area. In most cases the new nuclear production organization was placed under a vice president or a superintendent of fossil fuel power, usually called the Production Department. After all, the utility was building a generating station to provide additional electrical power the same as the fossil units. It had operators, mechanics, technicians, water chemistry, etc. just like the fossil plants. In-house up-and-comers were selected from the fossil plants to staff the cadre of the new organization. Non in-house expertise was hired from the military or was lured away from the Bechtels, Stone and Websters, Combustion Engineering, General Electric, etc. or was obtained from maritime academies.

Budgets were prepared and submitted to the Production Department executives who groaned and commenced to realize that nuclear power construction/operation was expensive. However, the dawn had not yet broken as to the fact that the new baby was about to abscond with the parents bankroll. This would come years later when Three Mile Island (TMI) sustained a mishap and the roof fell in on the capital improvements budget of the fossil fuel executives. Before that time, and after commercial operation was achieved, the fossil fuel executives displayed their fossil fuel mentality. "Why should I approve these outlandish requests for preliminary engineering calculations for a modification I neither understand, recognize as required nor wish to implement?" "The plant is operating, it's making money and my job is to reduce operating and maintenance (O&M) costs which are exorbitant relative to our fossil units." More than once it was brought to the author's attention by senior management that the NRC did not run this nuclear unit, the "company" did and it did not intend to spend unnecessary company funds on these frivolous modifications which do not satisfy cost/ benefit analyses. It didn't matter if we had, in-house, a bulletin from the NRC asking for our plans and schedule for complying with some problem encountered at another nuke, it wasn't even our design class.

How many times has this author heard the phrase "you guys in nuclear are going to have to toe-the-line and get in step with the rest of the company"?

The fossil fuel executives did more to hinder the growth in their nuclear organization than they ever realized until both shoes finally fell and the NRC ordered the plant shut down until further notice. It became standard practice to disallow staffing increases submitted with annual budget requirements. In some cases, reductions-in-forces was the battle cry of the utility while the nuclear organization was asking for staffing increases to do the "paperwork" required for nuclear compliance. The fossil fuel executives did not understand that compliance, as performed by the Inspection and Enforcement (I&E) branch of the NRC was to review documentation to ascertain proper performance and documentation required multitudes of personnel. The nuclear organization managers finally figured out how to obtain the necessary staffing to adequately perform their responsibilities. They would identify, to the regulator, the need for additional personnel before they could commit to a plan, schedule and completion date to comply with the regulators requests. At the exit meeting after an inspection and fact-finding visit, the regulators, and also in the regulators inspection reports, the NRC would report that the utility was deficient in certain areas and would request that the company provide plans and schedules for compliance. These requests were in the form of requesting written responses from the company. The nuclear managers would then submit their plans/schedules and staffing requirements to meet compliance. The fossil fuel executives, in signing out the response correspondence, were committing to the plans and schedules **and staffing** to implement the programs for compliance. These same executives were the ones who previously, at budget reviews, had rejected staffing increases from the nuclear supervisory personnel. Because it was identified from an external source and it was beholden of them that their signatures on outgoing correspondence committed the company to its word that they begrudgingly approved increased staff despite it being totally against corporate policy at budget time. The mentality in the nuclear regulatory arena rewards utilities who identify internal shortcomings before the regulator identifies them.

The fossil fuel mentality came to grips with the cost of doing business in the nuclear environment post Three Mile Island (TMI). The regulator started to issue Regulatory Guides which eventually cost each nuclear utility many millions of dollars to implement, including a vastly expanded Emergency Preparedness Program with drills, exercises, evacuations, etc, which cascaded all the way down to school bus drivers. Human Factors Engineering came into vogue with the requirement to reevaluate and modify control rooms after giving due consideration to present day human factors requirements. Post Accident Sampling Systems for sampling for radioactive gases and liquids were required to be designed and installed as a result of TMI. Additional and/or revised redundancy in instrumentation was mandated and instituted. The cable spreading room fire at the Browns Ferry Nuclear Power Plant in Alabama resulted in Congress enacting Appendix R to the Code of Federal Regulations and utilities were required to institute millions of dollars of modifications for safe power plant shutdown predicated upon a fire anywhere in the facility, including the control room, which is the nerve center for all power plant operations. Only after successive experiences with less than full compliance of the regulators desires did the fossil fuel mentality executives come to realize that the ante in the nuclear power plant poker game keeps going up because what was acceptable yesterday is unacceptable today. The target kept moving and rising and the effort had to increase to sustain adequacy. Future chapters will reintroduce the need for the utility executives to ante up to stay in the game.

Chapter 2

Nuclear Vs. Fossil—the In-House Manpower War

In the early 1960's this utilities corporate executives considered the various fuels available to power a needed additional power generation facility. Nuclear fuel was relatively inexpensive and steered the company away from further fossil fuel (oil) dependency. Manpower requirements were understood to be somewhat greater for a nuclear facility but were not deemed overly excessive. Nuclear was the wave of the future so their analyses resulted in the "in-thing", i.e., a nuclear power plant for the necessary generation requirements. An order for the design and construction of such a facility was executed and away-we-go!

After all, the manpower for such a facility was a minor cost of doing business. A fossil power generation unit required the usual, a plant manager, assistant plant managers to cover the 24 hour per day operation, operation engineers(watch engineers), operations shift supervisors, operators, maintenance supervisors and electrical and mechanical maintenancemen along with instrument, control, power and mechanical technicians as well as watch electricians, water quality technicians and janitorial and security services to name the cadre of manpower requirements. Note that all the listed positions are "hands-on" manpower. They operate and maintain the facility. A complement of administrative personnel round out the staffing.

The executives realized that the nuclear facility would require several additional disciplines, i.e. health physics personnel, quality assurance and quality control personnel as well as training and medical staff personnel, to name a few. So far so good!

Now let's move ahead in time to plant operations and see how staffing requirements increased. The plant is operating with "hands-on" personnel doing the responsible activities. We have added a couple of ALARA engineers (they make sure the plant personnel are exposed to as little as possible radiation absorption) when performing work in a radioactive environment. They recommend the installation of shielding and/or decontamination where possible. Lest we forget we also added a Regulatory Affairs Group to interface with the federal regulators.

There were approximately 120 nuclear power plants operating in the USA as well as many in foreign countries. Many foreign nukes were designed and constructed by US companies so their operation is similar to US plants. A requirement of the regulator is to notify the regulator of any unusual occurrences, be they equipment, operations, maintenance, human error, etc. Each reportable event is then written up as an I&E Bulletin (Inspection & Enforcement) by the NRC and sent to all nuclear facilities for review to determine if such an occurrence has any implications at their facility. One can compare this to an automobile recall for an inspection and/or repair.

Let's follow the NRC's issuance of a bulletin and its impact on the utility. Regulatory Affairs(also known as Licensing) gets the bulletin, logs it into the regulatory correspondence file system and takes it to the plant. Lets say it involves a faulty electrical relay operation which then resulted in equipment failure at the reporting facility. The Reg. Affairs guy meets with the electrical maintenance engineer and they discuss the bulletin. Now the maintenance engineer's job is to manage the routine and break-down maintenance activities scheduled to be performed. The bulletin is not on his "to-do list" and is not a high priority at the moment. Unexpected break-down maintenance may cause routine maintenance to slip. He says. "Yeah, OK. Put it on my desk and I'll get to it." What he is really saying is "someday, when I'm caught up I'll look it over." Several bulletins on his desk leads him to go to his boss and ask for additional help with the overflow of work. The request goes up the line and combined with other requests for manpower drives the executives to first stall and say "we will consider it at budget approval time" or get rejected outright or in some instances temporary help may be assigned.

The Regulation Affairs personnel go back to their desks and compose a letter to the NRC which states that "we will provide you with our plan and schedule for responding to this bulletin in ninety days." The letter triggers the NRC I&E Branch to put a tickler in their correspondence file. Ninety days goes by and the utility, not having fully addressed the bulletin, requests an extension. I&E puts the bulletin's status on it's agenda for discussion with the plant staff on the next site inspection. The electrical maintenance engineer, up to his proverbial eyeballs with a staggering workload, if not provided with addition assistance, lets the bulletin sit!

Comes inspection time and the I&E hoard are at the gate. They bring up the issue of the bulletin and the plant manager confers with his maintenance personnel who tell him they have not finished their review. The NRC carries this bulletin as an open item and notifies the utility senior management in its inspection report that the facilities response to various NRC requests are unsatisfactory.

More bulletins arrive along with the need to correct any I&E findings of deficiencies during its inspection and the workload increases. Pretty soon the utilities star starts to tarnish in the eyes of the regulator.

The fossil fuel minded executives still have as a corporate goal to reduce expenses and manpower, while the nuclear stepchild is requesting just the opposite, more money and more manpower.

Fossil fuel stations do not, as a rule, receive external events workloads that the nuclear facilities have and the fossil fuel minded executives failed to grasp that such conditions exist. It may well be that the nuclear managers were not sufficiently expressive to corporate that "that's the way it is down here."

After a while, corporate gets the feeling that the nuclear organizations performance is substandard and it is probably due to "lack of attention to detail." They are partly correct. After all the company has approximately 4000 employees and the nuclear organization has manpower greatly in excess of its fossil fuel stations. True, the nuclear organization is performing at a substandard level because the "hands-on" personnel do not have the resources to handle the extraneous workload. There comes a time when the manpower required to operate the facility in accordance with all the regulatory requirements exceeds that of the "hands-on" personnel.

Meanwhile the executives failed to realize that they were equally at fault for not supplying the manpower necessary to man the oars to row the boat upriver in the nuclear rat race! I call this the manpower war. The nuclear organizations workload varied in part due to external events beyond their control. A vice president of nuclear operations, who touted the corporate line of "less is more" when it came to staffing increases, was a victim of his own beliefs. He was dismissed.

Eventually the NRC, whom is often compared to the Registry of Motor Vehicles because they license and regulate, pulls you over and asks for your license and car keys. Your vehicle is unsafe to operate because it has not had all its recalls addressed.

Throughout this novel there are several instances where manpower needs became crucial to successful operation. Lack thereof resulted in instances of manpower burnout. Several chapters are devoted to identifying the need for new or modified work practices which then resulted in staffing increases. These fall under the executive aegis of "work smarter, not harder."

Chapter 3

How Do I Get into This Fortress Anyway?

Prior to performing work in a nuclear power plant one must hurdle the gates of entry requirements before one is allowed unescorted access within the security perimeter (the double fence surrounding the complex). This applies to all personnel who wish to enter, bar none. The mandate is there must be protection against sabotage from without and from within.

A non-utility contractor employee who desires to perform work must prepare for at least three days of preparatory screening and testing prior to admission. A contractor must attend several days of General Employee Training (GET) which covers all the necessary facets of the requirements for performing work, including, but not limited to: radiological protection, security, federal law, fitness-for-duty (FFD), right-to-know law, emergency preparedness (various alarms, sirens activated for various plant conditions), evacuation, radiation exposure, radiation work permitting, proper anticontamination clothing (anti-c), the proper procedure for suiting up and unsuiting, visitor requirements and one must pass a series of written examinations or be rejected from doing work at the facility. After successfully passing all the written exams, one must obtain a security badge and ID number with photo. The employees employer must submit a "good-guy" letter addressing the character of his employee. Screening for background will be conducted in a timely manner (within weeks). An employee/contractor must be free from drug and alcohol use and pass a physical examination prior to performing work at the facility. (More on fitness-for-duty in a future chapter).

Non-adherence to all the requirements for conduct when on the site are reasons for reprimand, suspension or termination depending upon the severity and frequency of occurrence.

Every entry into the facility requires screening for metals (guns, knives), passing through a bomb detector portal, request ones personnel ID badge, acceptance by the security computer and passage through a turnstile. Now you are aboard, just try to do some work!

Security News Item:

Again if one searches the Internet there are a multitude of items relative to nuclear facility security. One incident, found on the Internet and also reported on a major television news broadcast a few years ago, involved the finding of hundreds of lost keys to security areas by government contractors working at government nuclear facilities. This is not just at one facility but at several facilities. These findings were reported and found to be of a long standing deficiency at the government facilities. One article describes that at Oak Ridge, a facility that reprocesses and refurbishes old nuclear warheads, greater than 200 keys to protected security areas were missing. Similar incidences were reported at labs in New Mexico and in California.

These nuclear facilities have high grade uranium materials on-site and require strict security for national defense reasons. Several statements include the fact that the government is not satisfied with this situation and an investigation will ensue to correct any deficiencies. There was mention that corrective actions would require going before Congress and requesting millions of dollars to correct the problem. Sounds like they need a plan and schedule!

Author's comments:

If this was a finding at a commercial reactor site, the utility would have to immediately telephone the NRC and follow up with a detailed letter of findings. The NRC would send out an investigative team to assess the problem, request the utility implement compensatory measures immediately and report when the problem will be corrected and actions taken to preclude recurrence. The compensatory measures would possibly include stationing an armed security guard at every door designated as having a lost key or installing a new lock. The security guards would be stationed at each suspect door round the clock. Nowhere in any Internet items were there mention of the need for immediate corrective actions nor compensatory measures instituted. These are nuclear weapons producing plants! It isn't as serious as a lost key at a nuclear power plant that produces electricity. Does one get the feeling there are dual standards in play here?

So you are now inside the "protected area" but not inside the plant yet. Your ID badge identifies you by name, by number and by photo and is also coded to allow you access to various parts of the plant, depending upon your profession. Administrative personnel, etc. who have no need to access vital areas of the plant are coded for access to level I areas only such as offices, cafeteria, warehouse, medical building, training building, etc. Those who have a need for access to level II have their security card so carded. This allows them access to the reactor building, turbine building, control room, etc. Only those few personnel who so require are coded for access to level III areas like the refuel floor. If one attempts to access a secure area that is beyond his/her approved level, they are denied access when they attempt to gain entry by inserting their security card in the door card reader. The security computer recognizes this attempted intrusion and a security guard is dispatched to investigate the attempted intrusion.

"Tailgating"—this term identifies an individual who gains temporary entry into a restricted area. He/she may pass through a doorway by following an approved worker through before the door closes and without inserting their security card in the card reader. Tailgaters are suspended from the site when identified as being in a restricted area illegally and are required to re-establish approval for re-entry from plant security management after completing several portions of the General Employee Training(GET).

Additional security measures include closed circuit televisions along the security perimeter fence lines, E-field sensors between security fences, armed security forces, concrete barriers along plant entry roadways to prevent speeding trucks/cars from ramming the security posts.

Security, both external and internal at nuclear power plants is vital.

Identified weaknesses by the regulator may result in increased frequency of compliance inspections and/or fines.

Chapter 4

Are You Kidding—Fitness for Duty (FFD)?

Federal regulations instituted on or about January 1990 require all personnel doing work at a commercial nuclear facility to submit to and be tested for abstinence from internal alcohol and drugs on a random basis. Depending upon the average number of people who work at these sites in a twenty four hour period, a proportionate number of personnel are randomly selected by the computer to undergo fitness-for-duty testing that day. Theoretically, in a one year period, all personnel will have been selected and tested. One may be selected and tested today and be reselected tomorrow, next week or next month for fitness-for-duty screening. Over the course of the year the work population is generally screened for fitness to perform work. The randomness of the selection process mandates that everyone be free from alcohol and drugs at all times or be subject to failure upon testing and removal from the work site. In addition, all personnel are trained in personnel behavior observance for substance abuse and anyone suspected of substance abuse, at any time, may be requested to undergo fitness-for-duty testing.

The testing consists of taking a breathalyzer test and if the blood alcohol level is above the NRC standard of 0.04, the individual is considered unfit for duty. Urine samples are collected and sent away to an independent laboratory for drug screening. If an individual's sample indicates a drug content, a second sample, collected during the original fitness-for-duty test and frozen as a backup sample, is submitted for drug screening. All testing is considered blind testing with only an individuals reference number identified until such time as positive results are obtained. Personnel found with alcohol abuse are requested to attend a rehabilitation clinic, at the expense of their employer, and may be reinstated after completion of the rehabilitation. These individuals are then tested on a more frequent non-random schedule for concurrence of rehabilitation. Persons found to have drug substance abuse are suspended, and the nuclear industry is notified that this individual is a drug user. Persons found dealing drugs are terminated and their names submitted to the local law enforcement agency as well as the NRC and the rest of the nuclear industry.

The commercial nuclear power business is too important and must have employees with acute attention-to-detail acumen at all times. Neither the industry nor the general public can afford to have substance abusers performing critical work at these facilities.

Chapter 5

Redundancy, Redundancy, Redundancy!

Your automobile has been designed with some redundancy in it. It has dual master brake cylinders, dual brake lights, dual head lights, etc. Visualize if you can your automobile with its internal combustion engine, its power train, its exhaust and emission system, its cooling system, its safety systems and its manual of operation.

Now consider the automobile, designed and constructed to the code of federal regulations and the quality assurance criteria imposed upon it by the federal government. First you would have four water temperature indicators, and they would be of various designs/manufacture to preclude common cause failure, i.e., one flaw which could disable all four indicators simultaneously.

You would have four fuel indicators, four oil level indicators, four tire pressure indicators, four electrical power level indicators, four batteries, four alternators, four radiators, four sets of brakes, etc. In other words every essential piece of equipment would be redundant unto itself four times. If one electrical power indicator reads low, the other three would be read to try to determine what is wrong. If a selected two out of four indicators read low electrical power, two-out-of-four trip circuits would trip and stop the car automatically with no human intervention. In other words the instrumentation would place your car in the safest mode of operation, which is stopped. You would have four airbags per passenger and each would be activated by four separate sensors.

The principle of operation is simple, when something is wrong and at least two redundant circuits observe this condition, the car stops automatically. Now let's consider the manual of operation. You would require a utility trailer attached to the vehicle to transport the procedures manuals. Every thing that operates would have a procedure on how to operate it. Every alarm, indicator light/signal would have a procedure on what it meant, how it functioned and what to look for and do if it alarmed or activated. Also you would have the original equipment manufacturers manual which you would use to evaluate what malfunctioned and how to repair it. You would be required to keep in touch with the equipment manufacturer/supplier annually to see if he has changed any components or manufacturing process such that it could impact on how you maintain and/or repair your equipment.

Another section of your utility trailer procedures bookcase would contain test procedures. These may require you do certain tests before you start your car. Others would require you test circuits/equipment hourly, daily, weekly, monthly, every three months, every six months, annually, etc. Now everything isn't really that bad. You can, in some instances because of redundancy of equipment, test while the car is in operation.

Let's now consider what happens when a piece of equipment fails to operate when required to do so, or fails while in operation. You would consult the proper procedure for operation and for failure to operate. You would then contact your qualified mechanic who would write up a repair order and then consult with his procedures (from your utility trailer) on how to troubleshoot and/or repair/replace the failed component using the manufacturers/vendors literature, which you were required to maintain current.

But before the mechanic works on your car a planner/estimator takes the repair order and prepares a package (work package) which tells the mechanic what procedure to use, what the vendor says to do, where you can find replacement parts in the warehouse and whether the spare parts are certified and have not exceeded their shelf life. Additionally what tools to use, how long the job should take and what equipment needs to be isolated to perform this job. After all this, and the quality control personnel sign off on the contents of your repair package, you can go to work to repair the failed component while a Q.C. inspector follows the job to ascertain if you are following the procedure verbatim. You might also have the luxury of either having a resident NRC Inspector or in some cases a visiting NRC inspector monitoring your performance. Now you analyze the failed component to try to determine why it failed and then notify the NRC regional office who in turn will issue a bulletin stating any utility which has said brand of equipment in their facility to check it for possible flaws and/or subsequent failure. If the failed equipment was classified as safety related, i.e. it was part of a system required to safely shut down the car, additional requirements would necessitate obtaining documentation that the replacement component had met all the requirements imposed upon the original equipment such as seismic qualifications, temperature and pressure qualifications, etc. This simplistic example was meant to bring to you the complexity of design, operation, testing and maintenance of critical nuclear plant components. When you have completed the maintenance, you must then pull out an appropriate test procedure (again from your utility trailer bookcase) to ascertain the replaced equipment performs satisfactorily and in addition, that you have not disabled any other affected equipment during the course of this maintenance.

Why is all this equipment, procedures, processes, testing necessary? Because the operators of a nuclear reactor must, at all times, assure the performance of the plant relative to the requirements of the law, with the mandate to protect the safety of the public as the first commandment. If you cannot assure compliance with the requirements of the law, then you must shut the plant down, no ifs, ands, or buts, and right now! If the redundant components observe a disparity among themselves relative to the norm, they automatically shut the plant down without operator intervention or assistance.

To provide redundancy for the sake of nuclear safety, to allow for testing while in operation and to provide defense-in-depth to nuclear operation, the architects of nuclear power plants developed various redundant designs. All provide for the requirements set forth by the law. Visualize, if you can, the hundreds of functions required to operate a power plant, then multiply these hundreds of functions by hundreds of pieces of equipment to perform these functions and then multiply this by the redundancy factor of four and you can begin to envision the complexity of operation/maintenance of a commercial nuclear facility. Every function has a procedure on how to operate, how to troubleshoot, how to fix it, how to test it, when to test it, what the manufacturer requires for maintenance, testing, replacement, etc.

"Paper is the main product of a nuclear power plant, electricity is a distant byproduct." Future chapters will discuss, in-depth, the paper trail requirements.

"Someone once said that if a commercial airliner, with all its redundancy built in, had to meet nuclear power plant redundancy, it would not be able to take off due to its excessive weight."

Chapter 6

Your '65 Ford Now Needs a Catalytic Converter

The second generation commercial nuclear generating plants were licensed, constructed and started up and operated during the 1970's and 1980's. As previously mentioned the mishap at TMI and the fire at Browns Ferry resulted in Regulatory Guides, Branch Technical Positions, I&E Bulletins and Notices and in some instances, appendices to the Code Of Federal Regulations, all of which caused all nuclear units to evaluate their facilities against new or expanded requirements and then to implement improvements/ modifications to the majority of the operating nukes in the USA. Facilities under construction were required to assess compliance to the requirements and or comply prior to receiving startup approval.

I equate some of these mandated requirements to an automobile owner who goes to his/her authorized inspection station for a new inspection certificate or receives a notice in the mail from his dealer that his/her '65 Ford does not meet present day emission standards. Further discussion indicates that to meet present day requirements a catalytic converter would solve this deficiency. The owner pleads that he is grandfathered because his/her auto was not manufactured with a catalytic converter and met the emission standards back in '65. Too bad! You also receive a notice that snow tires or spikes are required during winter months, but your car is operated exclusively in Florida. Too bad! Next a notice is issued to examine your car to determine whether you have drum or disc brakes. If you have drum brakes you must upgrade your car to meet present day stopping standards, drum brakes presently cannot comply. Too bad! Next a notice arrives that loss of a battery or alternator/generator could cause power steering to fail to operate safely. Please inspect your car and notify us when you will make modifications which will correct this loss of battery power/power steering problem. Too bad!

This inundation of notices results in an inordinate expenditure of funds to bring your '65 Ford to 1990's standards. Welcome to the nuclear power generation upgrade game! In the nuclear generation game you must conform to present day standards. Case in point is Yankee Atomic in Rowe, Massachusetts. It did not meet present day requirements. A cost benefit analysis determined that it was not cost efficient to modify so it was shut down permanently. Yankee Atomic had two options: comply with present day requirements to the tune of multi-million dollar modifications or shut down. The cost/benefit analysis opted for shutdown. Public utility executives are inclined to approve modifications which either replace obsolete/expended equipment so as to maintain the generation capacity of the plant as designed or to expend monies to increase plant capacity and thus increase revenue. To expend monies to bring a facility up to present day standards with no return-on-investment (ROI) is contrary to the fossil fuel mentality previously discussed. Yet these executives were called upon to approve capital expenditures in the hundreds of millions of dollars to bring their 1960's designs up to present day standards with no ROI.

The 1970's and 1980's designs required the utilities to evaluate compliance for high energy pipe breaks versus the ability for safe shutdown. Additionally, the requirement to demonstrate shutdown predicated upon a fire in any area of the plant including the control room. Alternate shutdown capability was required upon loss of the control room. The plant must demonstrate fire detection, fire suppression, smoke ejection, discharged fire water removal and the ability to avoid electrical shock while fighting an electrical fire with water. The inability to meet the stringent requirements required modifications for compliance.

Cement block walls were evaluated for structural integrity during and after a seismic event. Piping system base plates and pipe hangers were evaluated for structural integrity relative to seismic events. The ultimate heat sink (the source of water for cooling pipe break steam leaks and for flooding the reactor core) were beefed up for increased structural integrity.

TMI's mishap resulted in the largest number of modifications/improvements. Boiling water reactors (BWR's) were not exempt from modification improvements despite the fact that TMI was a pressurized water reactor (PWR) with its inherent operating characteristics. Emergency Preparedness was elevated to new heights and the Federal Emergency Management Agency (FEMA) was empowered to evaluate and rate offsite emergency preparedness while the NRC continued to evaluate and rate onsite emergency preparedness.

Modifications to second generation nuclear plants have resulted in capital expenditures far in excess of the original cost of construction.

Emergency Offsite Facilities (EOF's), Technical Support Center (TSC), Operations Support Center (OSC) were all constructed to implement state-of-the-art emergency preparedness facilities with redundant control room information available at the TSC.

Security requirements were upgraded per the requirement of 10CFR50:73 whereby personnel accountability was mandated in all vital areas and within the security perimeter. Redundant computers track personnel accountability when onsite as described in Chapter 3.

Chapter 7

Forty Years Old and Just Like Brand New

"Your 40 year old car must perform like the day you bought it new."

    Safety related equipment, i.e. that equipment which must perform its intended function when called upon to do so to safely shut the reactor down and maintain it in a cold state is environmentally and seismically qualified to assure its performance. Nuclear generating plants are licensed for forty years of commercial operation. This forty years is forty calendar years and it does not extend beyond forty years if the plant were not in operation for several years.

    The license defines when it expires, when issued. Previously, in the chapter on Redundancy it was expressed that equipment is constantly tested based upon prescribed test cycles mandated by the NRC and are part of the nuclear plants license. Testing provides a high degree of confidence that the equipment will perform its intended function if and when it is required to do so. Testing also exacerbates the wearout of the equipment. We all realize that most equipment is not built to perform for forty years without wearout or failure. Therefore, safety related equipment has a useful life based upon the durability of the individual components which comprise that piece of equipment. Individual piece parts of this equipment has prescribed useful life based upon component qualification testing by the manufacturer. In order to ascertain a piece of equipment's availability for the life of the plant, components which have a useful life of less than forty years must be replaced either upon failure while testing, failure during operation or upon reaching the manufactures published useful life. As such, during preventive or corrective maintenance, equipment is disassembled and components replaced to bring the equipment up to a new condition. Components in this category can vary all the way from insulation, springs, coils, relays, storage batteries, fan belts, up to electric motors, circuit breakers, transformers, motor operated valves, and others.

As a result of component/equipment replacement and prescribed testing frequencies the nuclear facility maintains an availability status as near to new as possible. This process goes on day in and day out for the entire licensed cycle of the plant. Original equipment manufacturers (OEM) do not always maintain an availability of either components or equipment. Some discontinue the line of equipment, some modify the manufacturing process. Thus the requirement to maintain the equipment in the "as-good-as-new" status is complicated during the life cycle of the plant. When original equipment is no longer maintained certified for replacement or becomes obsolete, the engineering department must locate suitable equipment for replacement. If it can be purchased but is not certified as environmentally qualified, then laboratory testing must be performed to derive the useful life of this replacement equipment. Test laboratories exist in various parts of the country to conduct environmental/seismic testing. This testing greatly adds to the cost of the replacement equipment. In some instances when useful life is very short, i.e. less than five years, the entire piece of equipment must be replaced upon an expedited frequency so as to maintain it qualified for service.

The requirement to maintain safety related equipment functional for the life of the plant increases the maintenance workload, the testing workload, the documentation workload and the overall cost of doing business. This issue of availability for the forty year licensed life of the plant is another facet of the workload associated with nuclear power and adds to the nuclear burnout syndrome.

Equipment/component replacement is documented in maintenance files and retained for the life of the plant. If a component is used in the plant in both safety and non-safety related equipment, either safety certified components are used as replacements for both classes of equipment or certified and non-certified components must be warehoused separately and not interchanged. For instance, lubricating oil, bought in fifty-five gallon drums may be purchased with certification. This oil may be used in safety related equipment and non-safety related equipment. However, if this oil is purchased without certification at reduced cost, it cannot be used as lubrication in safety related equipment. The maintenance work package must specify if certified oil is required for a specific job or not. The quality assurance personnel would monitor the job and review the work package before, during and after completion for compliance to requirements of the job. This is one defense-in-depth factor associated with nuclear power. As different aspects of these requirements arise, remember that everything needs paperwork.

The documentation must be retrievable at all times to demonstrate to the NRC that conformance to and compliance with regulations has been maintained.

The author will not, at this time, try to describe the turmoil a facility must go through when it receives notice from the regulator that either another nuclear facility or the equipment manufacturer has deemed that certain equipment has not performed to prescribed and certified documentation and therefore a rigorous search must commence immediately to ascertain whether your facility is affected by this notice. If so, then off we go on another frantic search for suitable/timely replacement equipment and another chunk of work is now added to the already overflowing plate of other essential "stuff to do".

Singe is now beginning to show on the human body!

Chapter 8

One Changes the Light Bulb, One Holds the Ladder and Sixteen Document the Job!

The title of this chapter may seem facetious, however it approaches reality. For instance, to perform a corrective maintenance activity such as replace the motor operator on a motor operated valve we will follow the process from start to finish. Somehow, whether because of operational failure, testing failure, expired useful life, notification of component failure at another nuclear facility or some other reason, the motor operator is to be replaced on a piece of safety related equipment, i.e. a safety related valve.

A maintenance request is prepared and identifies the component equipment number, the name of the component, the system in which the component resides, etc. A work prioritization team reviews the maintenance request for necessity, plant conditions required to implement the work and prioritizes this job relative to all other requested/required jobs to be performed. The job is placed in a future window of opportunity for implementation relative to plant conditions.

The package then goes to the planning/estimating group who prepares the detailed work package including vendor information, steps to perform the job, procedures to isolate, test and operate the equipment, replacement component availability in the warehouse, tooling required to perform the job, etc. The quality assurance personnel review the package for content, quality, necessary inspection hold points, etc. and sign off on the package. The package is scheduled for implementation. Operations personnel isolate the equipment predicated upon plant conditions/needs. Maintenance personnel perform the replacement (change the light bulb), quality control personnel /inspectors monitor the job (hold the ladder), and the package is implemented and closed out. We have the work prioritization team (1), the planner/estimator (2), the work scheduler (3), operations (4), testing (5), documentation review for completeness/adequacy (6), document control who insert the document package in records management files (7), the maintenance clerk (8) who records the replacement equipment in the maintenance files. There are countless other personnel involved with this job including the maintenance supervisor (9) who reviews to Vendor Technical Information for accuracy and timeliness, the radiological protection technician (10) who surveys the area for radiation levels and prescribed stay time in the area, the ALARA (as low as reasonably achievable) technician (11) who reviews the job for ways to reduce radiation exposure to the workers, the Licensing engineer (12) who prepares the monthly status report of the plants operations record and major maintenance during the past month, the "Bean Counters" (13) (one of my favorite titles) who review the number of open, closed and completed maintenance requests each month and prepares a report for plant management, the technical services engineer (14) who prepares a report on the motor operators failure and submits it to the Institute of Nuclear Power Operations (INPO) for retention in the Nuclear Plant Reliability Data System (NPRDS) for all nuclear facilities. There will be detailed information on INPO and NPRDS in a future chapter. One must also include the administrative assistants to the operation department (14) who process the maintenance request and tagging/isolation package and the clerk/typist (15) who retyped the procedure (s) affected as a result of this work.

There are probably other personnel involved with the documentation flow but one can readily see that it takes a multitude of personnel to support the chap who actually does the job. It was not always this way, but every time the regulator finds a deficiency in the control of work i.e. lack of documentary proof of all that was accomplished, the utility responds by committing to provide additional proof, and bingo! bodies are added to the work process and staffing numbers grow.

Can you now begin to understand how the fossil fuel mentality was obsolete relative to nuclear power plant operation? In fossil: "We just go out and change the motor operator and away we go." "What's the nonsense of all this paper!"

Chapter 9

So You Just Made the Bad Guy List Huh!

Previously the design, operation, maintenance and modifications to a nuclear power reactor was discussed. Insight was also gained into entry requirements, fitness for duty testing, maintaining the equipment in a new, ready state and followed the documentation trail of one job.

Let's now consider the big picture. Day in and day out, equipment degrades and or fails, procedures become obsolete because of modifications, changes in operating practices, etc., manufacturers go out of business or quit making the parts for your plant. Personnel make recommendations on "how to" or "what to do" such as "how to do something differently or "what to" modify/replace/improve to do the job more efficiently and management's job is to juggle all these balls, select those with merit and benefit and assign people to go do them. All companies who produce something physical operate this way.

But there is another facet of the equation we must also consider. There were 109+ other nukes in the USA and they all have their problems. There are hundreds of nukes in the world and they all have their problems. Now, not only must management keep its own facility up-to-snuff, it must also evaluate, at the request of the regulator(s) or the manufacturers/vendors of plant equipment, how a failure or problem at another facility impacts upon its own facility. We have discussed Three Mile Islands problem and Brown Ferrys fire which caused all utilities to expend millions upon millions of dollars to negate mishaps similar to those at those facilities. There are also everyday small issues which crop up to divert immediate attention and the regulators arrive at the facilities gate, day-in-and-day-out, with "did you evaluate the impact of this occurrence at Plant X relative to your facility and did you document your conclusions?" "Were they positive or negative?" "Are you required by your evaluation to do something?" and "When will you have a plan and schedule for implementation?" Notice the magic words from the regulator; documentation, plan and schedule, These insignificant words have the hidden connotation of MORE WORK, MORE PEOPLE! These issues multiply and shortly, based upon input from you maintenance backlog, issues identified from other utilities, items identified from manufacturers, deficiencies from test procedure results, issues identified as deficient by internal QA/QC personnel. you have what is called a "backlog" of work and/or evaluation issues on your plate. The various regulators who visit the facility, sometimes announced and sometimes unannounced, as well as the in-house inspectors, evaluate your performance on "addressing the issues", all the issues, and keep tabs on how you are doing.

With too much work and not enough time or personnel to address all the issues, human nature being what it is, tends to cause management to address significant issues and disregard or prioritize to a level where certain issues do not get addressed. Utility management must balance productivity i.e., power generation versus regulatory issues. Utility managers are measured internally by corporate management for what they deliver to the corporate coffers while regulators measure the utility as to how well it performs relative to the laws and all those outstanding items the regulator keeps asking what the utility is going to do about. The federal regulator does not care whether you generate any power, its responsibility is to protect the general public and does this by assuring itself that commercial nuclear power reactors comply with all the laws and regulations set forth by the federal government.

In several states, utility managers who operate commercial nuclear reactors are measured against two sets of standards and in some cases these may diverge. For example, Public Utility Regulators establish annual goals for which utilities are rewarded or are penalized if these goals are not achieved. These goals may be contrary to the federal government requirements for operation. Generation is a major goal along with capacity factor and availability to satisfy state public utility regulators whereas the federal government may deem the operation of a facility which is not in conformance to regulations unsafe and require shutdown which then penalizes the utility against generation goals.

Often times, and when due diligence over a long period of time is not maintained, a facility gets a large backlog of work, equipment failures may cause a reduction in power or shutdown and in general "things are not going well". In due time the regulator comes along and says your facility is not performing to the desired standards and therefore the plant is placed on a "watch list" which plant personnel call the "bad guy list". This mandates increased attention by the regulator and puts all the activities at the facility under the microscope. You are criticized by the general public, other regulators join the bandwagon because the facility is not satisfying their requirements either and the utility finds itself sliding exponentially toward hell!

"Your intentions when joining this outfit was to drain the swamp. No one said you would be up to your ass in alligators."

Chapter 10

Rising Standards of Excellence or You Better Row Harder

For the moment envision a work place where you never receive a pat on the back, never a word of encouragement, like "nice job", and satisfactory is the highest grade you can attain. Rather cold and foreboding isn't it?

The federal regulators grade all US nuclear facilities once a year in six (6) disciplines: operations, maintenance, engineering, radiological protection and control, security and training. Quality assurance and quality control are graded across all categories, not individually. The highest(best performance) grade is category I which signifies "satisfactory performance."

Category II denotes "requires improvement" and Category III denotes "requires escalated attention by the utility and the regulator."

Let's for example say your utility received four category I's, one category II and one category III as your latest report card for your performance for the past year as viewed by the regulator. You would assume that the four category I areas of operation were performing at a satisfactory level and required minimal attention to detail. Your efforts would naturally be directed toward the category II and III areas which need improvement and should get your support. WRONG! What is satisfactory today will not be satisfactory next year. The magic words to describe this are "rising standards of excellence."

Those four words, along with the infamous "plans and schedules for implementation" denote an increase in workload, an increase in expenses and an increase in upper management involvement and commitment just to maintain the status quo.

No wonder the silent nuclear syndrome of human burnout prevails. You can never let your guard down, you can never rest on your laurels and can never enjoy the feeling of complacency.

To the best of the authors recollection, during the twenty-seven years of nuclear operations involvement, there have been at least thirteen plant managers. Some lasted four or more years, one lasted one week. The one week candidate saw the writing on the nuclear wall and said "thanks, but no thanks." Thirteen plant managers in twenty-seven years equates on average to 2.07 years on the job as the fall guy. The pity of it all is there is pressure from within (executive management), there is pressure from several external sources (regulators). There is pressure from your own QA/QC organization and there is no perceived reward with the exception of your salary. You are in a no-holds-barred poker game, your resources are limited and the other players keep upping the ante. Burnout is inevitable. I have seen ideal marriages fail, increased medical problems and I have seen dedicated performers kick it all because the stress was just too great to bear.

Every incident at the facility must be reported to the NRC, no matter whether it is significant on insignificant and this information is available to the press. Articles about the facility only increase the general public concerns and resentment toward the nuclear power industry.

One important aspect about rising standards of excellence is that nuclear facilities are graded and then compared to sister utilities. Not everyone can receive all category I results continuously. The NRC Inspection & Enforcement Branch would have nothing to do nor anywhere to go if all the nukes were AOK. So somebody has to be the bad guy and over the course of many years, at one time or another, all the nukes have been at the bottom of the barrel. Human nature can only persist for so long when there are several mountain ranges yet to climb beyond the peak you are presently scaling. The tenor of the job becomes oppressive when the positives are few and the negative aspects become overwhelming.

You probably heard the quip about the galley slave master who tells the poor souls condemned to rowing the ship "rest today because tomorrow the captain wants to go water skiing." That's how the nuclear game is played. The regulator is the captain.

It would be negligent if at this point the nuclear industry and the regulators in the USA were not complemented. There has never been a nuclear fatality in the utility industry. There is a standard quote which nuclear personnel in the USA cite when discussing the safety of nuclear power. "More people have died in the back of Ted Kennedy's Oldsmobile than have died as a result of nuclear power plant operations."

Chapter 11

Too Many Masters and Not Enough Slaves

Previous chapters and the foreword have described those entities who were, in essence, masters in the nuclear power generation game. The description of master is relegated to all those entities who demand something from the plant staff whether they be regulator, inspector or owner. Some of those masters are:

-Executive Management
-NRC with several different branches
-INPO—the Institute for Nuclear Power Operations
-State Public Utility Department
-Quality Assurance/Quality Control at the utility
-Local and County Selectmen or Commissioners
-Environmental Protection Agency
-OSHA—Occupational Safety and Health Agency
-On-site NRC resident inspectors
-FEMA—Federal Emergency Management Agency

Masters responsibilities vary and in many instances are diverse in nature. All request responses, investigations, evaluations, and commitments to perceived or actual deficiencies. Previously we discussed the diversity between the state public utility department/ agency and the federal nuclear regulators. Findings, as a result of actual inspections or utility self-identified deficiencies, require written responses from the utility, concurring with the findings and committing to corrective actions to preclude recurrence, along with those dreaded words—plans and schedules for implementation. Thus there developed numerous, independent deficiency lists which eventually, for the sake of cohesion, sanity and the only way to determine how big the commitment backlog was, were integrated into one huge backlog of work. The individual action items were maintained by the those expertise's responsible for implementation but were subsets of the master backlog database. Numerous items were interrelated and they required integration so as to avoid duplication or cancellation of commitments when one item affected the results of another item. For example, several changes to the same operating or maintenance procedure were identified as a result of numerous independent commitments. Thus the procedure was identified as requiring revisions in several data base items. Someone was responsible for determining that the procedure changes did not affect each other to the point of becoming another deficiency in and of itself. Databases included INPO action items and backlog list, NRC (several) action items and backlog lists, QA/QC deficiencies items and backlog lists to name but a few. All action items/deficiencies required assignment for implementation (ownership), corrective actions required to derive closure, manpower assignments including integration across various disciplines, scheduling if the corrective action required physical changes to the plant configuration, documentation updates, procedure revisions, equipment database updating, drawing revisions and if necessary, plant personnel training if the corrective action resulted in changes to operating equipment. Is it becoming clearer that every glitch was a major work effort?

The increased work scope required increases in manpower and increases in either capital or expense budgeting. Thus we always had a shortage of "slaves." There seldom was a budget request review that did not involve the request for manpower increase and the rejection of the increase in forces by executive management. One can still hear the responses advanced by executive management:

-work smarter, not harder
-we provided you with computers to manage the databases

-the corporate goals are to reduce company staffing
-the inclusion of this work does not provide a cost/benefit to the corporation
-there is not return-on-investment for these expenditures

One can picture the Sr. V.P.Nuclear at the budget review meetings at corporate headquarters biting his tongue when all other Veeps are submitting budgets with no increases or reduced staffing and the Nuke Veep has a request from his organization for umpteen additional people.

One such review with a senior vice president led to the response that the benefit was "allowing continued operation of the facility by the regulator." That response was not found to be humorous. Again, the two means of fighting the war on both fronts, getting the resources to do the work and getting the regulator off your case were (1) you had the regulator identify the manpower shortfalls to implement the plans and schedules advanced by the onsite personnel and (2) you had the corporate executives sign off on the commitment response letters thus insuring, under their signature, that you would receive the necessary manpower and funding.

There were times during the operation of the plant that deficiency resolution/correction outmanned, outspent and was a bigger work force than plant operations, maintenance and testing combined.

The Institute for Nuclear Power Operation(INPO)

Several times the mention of INPO was cited as a regulator. Here is why I designate it as a regulator.

After TMI, the collective input from all public utilties with nuclear power genertion deemed it essential that the industry institute, staff, fund and respond to an organization which represented the industry and was tasked with quasi-regulatory power in as much as it inspected all aspects of the nuclear facilities operation and identified weaknesses or outright deficiencies relative to satisfactory performance. INPO provided reports and inspection findings at all the nuclear facilities. It came to pass that INPO became another regulator simply by the fact that its findings were reviewed and incorporated into NRC inspection observations to determine if the facility had implemented corrective actions to INPO findings. The purpose for establishing INPO was to allow the utilities to identify deficiencies or performance below NRC requirements and to take corrective measures before the NRC discovered them. The NRC used these findings as a stepping off point from which to inspect the utilities for conformance.

The slaves were up to there proverbial eyes in backlog item reduction. Personnel were assigned to provide senior management with backlog reduction performance reports(more bean counters).

A later chapter is devoted to the fate of a nuclear operation vice president who did not take INPO seriously.

Chapter 12

Mess with the Bull and You Get the Horns

Sometime along about the tenth year of operation of the nuclear facility executive management decided to place the nuclear organization under a dedicated senior vice president for nuclear operations with its own budget, staffing complement and corporate goals. Previously a vice president-nuclear reported to a senior vice president of production or operations, whichever was in vogue. The new senior vice president-nuclear decided to broaden the management oversight by having two vice presidents report to him, one for engineering, quality assurance/quality control and emergency preparedness. The second vice president oversaw operations, maintenance, radiological control, nuclear training and operations support. The Operations Support Department then became the dumping ground for all activities not directly related to either engineering or operations and became home for all the homeless. At one point in time seventeen people reported directly to the Operations Support Department Manager and they included the division managers of Records Management/Document Control, Regulatory Affairs (Licensing), Environmental Sciences, Security, Plant Technical Support and twelve Capital Project Managers for plant modifications/improvements as a result of TMI, Browns Ferry and other identified safety concerns.

The Vice President Engineering was already in place, having held this responsibility from day one. The Senior Vice President-nuclear interviewed and hired a Vice President-Operations and Operations Support. The individual hired previously held a middle management position in another nuclear facility, never having held executive management responsibility.

This individual had high ambitions, understood the corporate goals and was bound and determined to meet or exceed these goals or die trying. One of his first declarations to his organization was that there would be reductions in staffing in his organization either by attrition (retirements, resignations) and/or via realignment of responsibilities to "work smarter, not harder." His second declaration was that henceforth there would be no so-called ALARA Engineers walking around. Everyone would be an ALARA Engineer and contribute to the successful reduction in job performance radiation exposure.

Next he proclaimed that he would know when we had achieved critical staffing i.e., the absolute correct number of personnel to do the work. "When we reach that critical number we will know it because one less worker would upset the organizations ability to perform its responsibilities." (I never could figure out what he was smoking!) Needless to say, his declarations went in one ear and out the other. Here was an individual who was trying to manage a workload of indeterminate size at any given time and he was going to know critical mass of employees, by instinct I guess!

Records Management/Document Control is that resource who have the responsibility of recording, documenting and controlling all paperwork whether it be memos, drawings, specifications, calculations, vendor manuals, procedures, or what have you. There came a time when the division manager made a presentation of the plan and schedule for backlog reduction to his superior, the Operations Support Department Manager. It was acceptable, workable and did not require any increase in staffing but required the expenditure of many thousands of dollars to hire outside expertise to do the work in-house, via contractors. The Operations Support Department Manager then had the proponent make the same presentation to the Vice President-Operations so he could recognize the magnitude of the problem, agree with the plans and schedule and request corporate funding so as to proceed. One must understand that, legally, all records management and control of documents are a safety function because the facility must, at all times, be cognizant of the latest plant configuration. Configuration control is part of the operating license of the plant. So documents not yet entered into the control system must be retrievable and all personnel must be aware of their existence. During this presentation the Vice President-Operations needed to make a phone call (to see if his new company car had arrived at the company garage)! This is the individual hired to run nuclear operations and his priorities were to see if his new company car was aboard? The Records Management Division Manager picked up his slides, his budget numbers and his presentation and walked out. Later that same day he submitted his resignation with the inscription across the top which said "he is an egotistical idiot!" Not surprisingly he went to work as a consultant to the contractor the utility finally hired to manage the records backlog. Other incidents occurred such as the vice president demanding that operations support personnel, located some 30 miles from the plant, work weekends, without overtime pay, because plant personnel often worked weekends. That request went unfulfilled as well. If he wanted to work weekends to do his job, all the more power to him! All this leads to the critical issue that became the undoing of the Vice President-Operations. The Institute for Power Operations (INPO) completed one of its annual inspections and issued a findings report citing numerous shortcomings, deficiencies and areas of concern relative to satisfactory performance. The Operations Support Division in the Operations Support Department had the responsibility to manage the disposition of the report by preparing responses advanced by and agreed to by the owners of the problem (area of responsibility). These responses required commitments to implement corrective actions to resolve the deficiencies such that they would preclude recurrence. In other words these were actions of substance, not band-aids. Several weeks later, with input from those disciplines responsible for implementation, the response report was circulated throughout the nuclear organization, primarily for adequacy of the responses and for concurrence that the actions would provide corrective measures so as to preclude recurrence. All sub-tier disciplines signed off denoting their approval. The responses required two further approvals, by the two vice presidents.

The Nuclear Veep of Operations requested a meeting on a Saturday morning to review the corporate responses. The meeting was staffed by the V.P. Nuke-Ops, the Ops. Support Dept. Mgr. and the two lead engineers who managed the work of obtaining input to write the responses. Almost immediately the Veep stated the responses required rework. We were not going to commit to such a large work scope. He then proceeded to dictate the revised responses which committed the company to either perform band-aids to the problem or word-engineer responses which denoted the company was reviewing the issues and is undecided at this time on how to correct the concerns. He revised the word "deficiency" to the word "issue". The loyal slaves proceeded to rewrite the report responses and had clerk typists prepare it with a transmittal cover letter to INPO. Less than one week later the Senior Vice President, Nuke was telephoned and was requested to travel to INPO in Atlanta, Ga. to discuss the response commitments. That fateful day was a Friday. The SR. V.P. was reamed a new rear end and told that his company was performing poorly relative to nuclear satisfactory standards and that obviously he wasn't serious about raising the standards to acceptable levels. That same Friday evening the Operations Support Manager received a phone call at home ordering his attendance tomorrow morning at the station managers office along with the station manager. The tenor of the conversation was such that it implied 'heads will roll."

Saturday morning with the two managers in attendance the Sr. V.P. Nuke indicated he was embarrassed and humiliated at INPO who stated that the nuclear facility was in trouble and didn't even recognize its poor performance. No notes were taken as the SR. Veep read the companies responses and moaned. The Veep of Operations wasn't at this meeting which caused some concern. Along about the sixteenth response the SR. Veep looked up from his notes and said, without skipping a beat, "the Vice President-Nuclear Operations, per my request, has submitted his resignation and is now terminated."

"I am considering several other terminations." And he then proceeded to complete his findings at INPO.

After the dressing down the SR. V.P. and the Operations Support Manager had a sit-down to discuss the "hows and whys" of this poor performance. Luckily the plant word processors retained all the previous report revisions. The original report, prior to the band-aid version, was brought forth and it was agreed that the original in-depth response report was to be submitted immediately under the signature of the SR. V.P. The request by the SR.V.P. to transfer the two lead engineers out of the nuclear organization was vehemently objected to as being needlessly vindictive. One engineer later voluntarily transferred back to the fossil organization. The Operations Vice President was gone come Monday morning. He had been terminated for under performance, not to corporate goals but to consideration of the health and safety of the general public! No one could recall an executive terminated for other than theft of corporate funds, sexual harassment, lying, malfeasance, or misfeasance in office. We all assumed he had found that his organization had one to many people on board.

It wasn't too much longer that the facility, overburdened with a backlog of work too big to manage, incurred the wrath of both INPO and the NRC. Plant operating abnormalities gave the regulator the opening to demand that the facility be shut down and not restarted until all outstanding items of concern were addressed and corrected. The corporation did not have the manpower, the money or the time to address the concerns while operating but commenced to staff, overstaff, commit, over-commit and fund millions of dollars and more than two years of shutdown to correct those concerns originally identified by the nuclear organization itself or by the external regulators. Lessons learned are hard lessons! Thirty three months after shutdown the NRC gave approval to restart under NRC scrutiny and with many new management staffing additions and changes. The Senior V.P. Nuclear returned to the fossil fuel organization. Another chapter will discuss the nitty-gritty of being on the bad guy list.

That being said, many messed with the bull and got the horns!

Chapter 13

Documentation Up the Ying Yang!

A nuclear reactor runs on enriched uranium so the story goes. Yeah right! It runs on documentation. You may have a refueled core with two years supply of fuel but, if you do not have complete documentation to assure the regulator that you are in compliance with configuration control, you may not be in a position to generate power. This has occurred on several occasions in the history of plant performance. A future chapter is devoted to configuration control.

Assume for the moment that there are approximately 200 personnel in the nuclear organization and each individual generates about ten documents per day. Documents would include, but not be limited to, descriptive memos, calculations, drawings, procedure pages, vendor manual revisions, communications from external sources relevant to plant configuration. There are many more sources of documentation, but you get the point, all of which comprise the design of the plant. Important telephone calls which result in action items and/or agreements for revising plant configuration would be written up and classified as components of configuration control.

One would not just up and change a procedure, they would be required to prepare a procedure change notice (PCN) which is assigned a number and tracked and becomes part of the documentation which must be retained for the life of the plant. One would not just up and change a drawing, they would prepare a drawing change notice (DCN) which is assigned a number and tracked and becomes part of the documentation likewise. All that being so, 200 individuals generating ten documents per day would become 2000 documents per day. External documents such as correspondence from the regulators, etc. also add to the documentation heap.

Now assume there are approximately twenty-five individuals in the Records Management/Document Control Division, four of which are assigned to process documentation prior to insertion into the document control system. Each document is scanned, read and then the title, document type, plant numerical system the document relates to and several keywords are assigned.

The document is then transferred to another individual who will microfiche it and place it in the controlled system. Four individuals doing the work, at approximately fifteen documents per hour (one would only wish) totals 480 documents inserted into the control system , such that anyone seeking to retrieve these documents can readily do so by requesting the system number, type document and some key words to describe what the document is all about. Non-retrieveability makes the records management system useless. So 480 documents are inserted in the system and there is still a backlog of 1520 documents for that one day which are still orphans. They are in limbo, neither here nor there. One can readily see how a backlog of documents accrues. Uncontrolled documents can and sometimes do result in less than full compliance with configuration control. This is a no no! By law, the facility must be in full compliance with configuration control. The regulator recognizes that everything can not be done at once but they demand controls such that compliance can be demonstrated in a timely manner.

The organization could not, nor would not, hire a bevy of clerks to process documentation. Instead the backlog is scoped and bids are requested from contractors to provide catch-up. This is cheaper and faster than internal means. There have been, at times, room(s) full of boxes containing documents awaiting incorporation into the documentation system. Nothing in this world is instantaneous so there is a delay associated with full compliance.

An individual performing work at a nuclear facility must have the full knowledge of the present plant status relative to its configuration. An operator must know how the equipment works. An engineer must know what outstanding changes exist to a drawing. When he/she requests a drawing from document control he/she receives the drawing plus all the outstanding drawing change notices (DCN's) to that drawing so as to be knowledgeable as to the existing plant configuration. It is not an easy way to do work and it could be fraught with pitfalls which could reflect poorly on the organizations performance in the eyes of the regulator if due diligence is not maintained.

This is but another chapter in the mountain of controls and management responsibilities one must perform to satisfy the regulator and protect the health and safety of the general public. Another straw on the camels back. When will it break?

Chapter 14

Configuration Control: Easy to Say but Tough to Implement!

Previously, the discussion on documentation cited the need for configuration control. What is configuration control? Let's step back to when the plant was constructed. The architectural engineering company and the nuclear steam supply company turned over to the owner of the utility all the documentation generated from the inception of the contract to build the power plant. The numerous systems (more than 60) all have a system number and all relative documents contain the appropriate system number to provide configuration control. This is how the plant was built, this is how it operates, and these documents reflect the present day configuration. The configuration at turnover is the initial configuration and the processes to maintain this documentation up-to-date is called configuration control. Any, I quote, any change to the initial configuration requires that all relevant documentation associated with any change must be updated, in a timely fashion, or the control system for the process must provide retrieveability of all not yet updated documents. That is the only way that configuration control may be maintained, as required by federal law relative to commercial nuclear power generation.

The regulator routinely inspects for configuration control. Every time they appear at the facility and request information to ascertain compliance, the documentation provided must demonstrate that the processes assure configuration control. Failure to do so would result in a deficiency and the seriousness of the offense would dictate what the regulator will accept as proof of compliance and that the facilities corrective actions will preclude recurrence.

What happens when a plant control system is revised and all the licensed operators and supervisors go to operate the system as they were original taught from startup? The control system operating features have been modified. Do you think this would cause some consternation when expected performance does not happen? Obviously there must be some management controls which allow all operators working three shifts per day, seven days per week, to get up to speed on what has happened and why. There must not be a situation where licensed personnel become confused during the normal operation of the facility. Rapid updated document dissemination and up-to-speed training must be provided before this becomes a problem. A future chapter entitled "The NRC turns the screws up a notch" will address this very issue.

Only by being a part of the process can one appreciate the magnitude of the efforts required to maintain configuration control.

Imagine you drop your car off at the dealers for routine maintenance and unbeknownst to you the dealer implements a small modification to your cars operating system because a bulletin from the manufacturer recommended the change to preclude a dangerous operating condition if the change is not made. He/she forgets to tell you about the change thinking you read about it in the newspaper or saw it on television. You drive off and suddenly it begins to rain. You reach for the windshield wiper control switch, turn it on and nothing happens. You try it again and still nothing happens. Are you perturbed? Did the mechanic blow a fuse? What have they done now to screw up your car?

You pull over and call the dealer who tells you that you must not only turn on the control knob but you must push it in at the same time. The dealer has changed the configuration of your vehicles' operating system and has not trained you on its operation. You and the dealer just messed around with configuration control. Not only that but the owners operating manual hasn't been changed to reflect the new condition so if you sell the car the new owner won't know how to operate it either. Get the point? Configuration control at a nuclear power plant is of the utmost importance for the safety of the operating personnel, the equipment and the general public.

Another straw on the camels back. The load is getting quite heavy and the prospect of relief is out of the question. It will only get worse due to rising standards of excellence. Burnout is on the horizon!

Chapter 15

The NRC Turns the Screws Up a Notch

A few years after the TMI accident the NRC, imposed, not by law but by sheer brute force, the imposition of plant reviews against Information Notices, Information Bulletins, Branch Technical Positions and probably several other "non-law" edicts. Plants were required to respond as to whether or not they could meet new stringent requirements with the systems presently installed. Naturally none could meet the new requirements and so the dreaded statement was "please provide your plans and schedule for compliance." In addition to TMI mandates, the Brown's Ferry fire and Appendix R to the code of Federal Regulations still was looming and the same questions/reviews/inspections and responses were required. All-in-all, the utilities probably had more than 12 projects on there plates and several were multi-faceted in scope. The utilities were inundated with the need to install millions of dollars of modifications to the existing facilities to comply with the new standards imposed by the regulator.

Architectural engineering firms such as Bechtel, Stone and Webster, Brown and Roote, etc. were literally awash with work to meet the demands of all the utilities to satisfy the regulator. It was estimated at one time that the major engineering firms would need to hire 25% of all graduating engineers in this country to satisfy the work load. Naturally as the work packages arrived at the site for implementation along comes Uncle Sam's Feds to oversee the work scope. The underlying tone from the regulator was that configuration control would be more stringent(rising standards of excellence). No more would it be part of the controlled backlog of work. It was imperative that, because of such a huge amount of modifications, the control of configuration would need to be completed so as to ensure that everyone and everything was current and up to speed.

Two middle management managers saw the handwriting on the wall. The Plant manager, responsible for getting the plant back on-line after a modification or refueling outage and the Operations Support manager, whose responsibilities included the projects, under the project managers and the construction management group who supervised the actual modification construction as well as the licensing group who communicated formally with the regulator. The need for expedited configuration control so as to shorten the downtime before startup was identified and discussed. The product of the meeting identified several documentation areas that needed close supervision. These included :

-procedure revisions prepared, approved and disseminated
-plant drawing revisions prepared, approved and disseminated
-vendor manuals revised or incorporated into the system as applicable
-preoperational tests prepared, approved, performed and documented
-plant personnel trained on the new modifications along with the operating, maintenance and testing procedures
-technical specifications revised, i.e. the specifications which describe how to operate the plant safely or shut it down(the license) and disseminated.

The list of documentation required to ascertain configuration control is not new or alien. The problem is that, in a matrix organizational structure individual owners of documentation preparation and approval requires the assistance of several individual disciplines to complete. For example the training department need the revised procedures, revised drawings and revised technical specifications in order to prepare lesson plans and train the operators, mechanics and technicians who operate, maintain and test the equipment and systems. The operating department needs system performance documentation in order to write operating procedures and so forth. This haphazard method of completing work in a timely manner is, at best, disjointed. One persons priority to support another persons priority may not coincide.

It was agreed upon by the two managers that a dedicated work control system should be instituted, for the benefit of the company, to manage the configuration control for major modifications. Raytheon Corporation was contacted and a work flow experienced individual was interviewed. Eventually Raytheon, who performs this type of work for the Federal Government, was hired to develop flow charts for all the major components of each documentation process. These were reviewed, revised and then used to develop work control procedures as well as the checklists and appendices required of each work control system. The entire product was then presented to executive management VP's along with staffing requirements and budget. Actually the cost of performing this type of work is capitalized along with the modifications. There was overall approval of the need and the solution to the need, i.e. a modification management system, whereas configuration control was a requirement for plant operation. The multitude of modifications made configuration control an enormous job. The staffing was a snag and all agreed the work would be capitalized and not affect the organizations operating and expense budget.

At about this time the facility was shut down to refuel and implement several plant modifications. The Operations Support manager drafted two individuals from his organization to temporarily staff the new group effort. Also two Raytheon engineers were contracted for as well as several independent contracted engineers and a temporary clerk/typist/secretary. A utility trailer was rented as office space, and away we go!

By the time the outage was winding to a close, the NRC inspectors appeared and inquired about the status of the major modifications. The new process documentation was retrieved and shown to the regulators. To make a long story short, the outage was extended by three days while the regulators reviewed the documentation which confirmed compliance with configuration control. Had the work control system not been in place and the need to revise procedures and drawings, insert vendor manuals, pre-operational test the new modifications and train the necessary personnel, the outage could well have been extended several months. The results sold the system to executive management who approved permanent staffing for this work process. Showing is believing!

Several months later a manager from the utility attended a seminar in Washington, DC where the topic was configuration control across numerous projects prior to startup. The majority of attendees were from facilities not yet granted commercial operating licenses. Therefore their experience was primarily to accept the constructors documentation for configuration control. The modification management work control system was presented and the attendees were overwhelmed by the need for them to develop some form of work control once they became commercial entities. All requested copies of the operating manual. All were directed to contact Raytheon to tailor there needs to their systems. This was one mountain that was scaled before the snowstorms hit, i.e. configuration control was in place at the same time the NRC was jacking up the ante!

Chapter 16

The Long Term Plan (LTP)—Sanity at Last!

In previous chapters there was mention several times of the multitude of projects ongoing and of the magnitude of these projects in scope and cost. Utilities were literally shoveling millions of dollars of improvement modifications into their nuclear facilities, primarily to satisfy the regulators demands for upgrades in safety in the unlikely event of a TMI type incident. Architectural engineering firms along with nuclear steam supply companies were inundated with work. Real concerns arose when there were questions asked like "How do we know one modification was not negating another modification?" "Is all the physical hardware being installed impacting safety in the name of safety?" "Who is stepping back and taking the big picture look at final product?" It was the responsibility of each utility to answer these questions and it was the responsibility of the regulator to review and approve the utilities actions. Therefore dual responsibilities existed. There was also mention of the fact that, as a result of all the capital expenditures required to meet upgrades in nuclear safety, the utilities did not have the means of investing in upgrades to increase return-on-investment (ROI).

Somewhere along the way, when the NRC noticed that the utilities could not meet their plans and schedules for implementation and completion of all the work, someone in the NRC Regulatory Branch began questioning whether the NRC, in its zeal for public safety had compromised that safety by the headlong crunch of making the utilities rush the modifications. Several visits to all the utilities resulted in the NRC getting the same responses nationwide, the sprint marathon was too much, too fast and too long.

The next question was, so now what? It was brought up that all the NRC issues were superseding all other utility desires. They also heard about the fact that there could possibly be safety concerns along the way relative to the big picture. Lo and behold the NRC said all right, let's work on an interface between the parties that was doable and workable simultaneously. That meant that the NRC had to back off of its demands for expedited results. Suddenly the light had been turned on to expose the overbearing pressures, the mental lapses and the crushing responsibilities that had been placed on the individuals in the nuclear industry by the regulator. All the work that was committed to, plus all the work not yet committed to were tabulated and present-day schedule commitments reevaluated.

It came to pass that the NRC, along with all the nukes, settled on a program of implementation that was rational to both entities. The Long Term Plan (LTP) was born. The main thrust was that the utilities would put all the work into a master plan and the issues classified. "Requiring expedited implementation" were categorized as class A items. Items which the NCR required to be implemented on a normal schedule" were classified class B items. Items the utility proposed to be funded and implemented were categorized class C items. The utility could, with the regulators approval, place a class C item to be addressed before a class B item. The Long Term Plan became an appendix to the Technical Specifications, the book on how to run the plant within the guidelines of the license. It was now a part of the operating license and was updated and sent to the NRC for review every six months. Updates contained progress and possibly reclassifications for approval.

There finally was a sense of relief, the marathon was now slowed down and extended. Work could be performed in a rational, safe fashion. Deadlines would be extended based upon sound reasoning. It was now the responsibility of the utilities to examine the in-house workload, agree upon selected improvements, manpower load all the work required to satisfy the plan and this produced a schedule for implementation with milestone projected dates. Some schedule end dates were not satisfactory to the regulator and resulted in discussions and or reclassification of other regulator desired work to new extended dates. The regulator had to give in order to get!

They could not now have everything they wanted, when they wanted it.

A software developer was contracted to develop the several desired computer screens for each work entry. The screens were for description of the project, classification and responsibilities. Other screens contained level I schedules and running cost tracking capabilities. At one time there were over 200 items in the Long Term Plan (LTP). Only a dozen or so were major projects of class A category. Many were for class B items and a few were for class C items.

The managers in the nuclear organization met once per month with the LTP administrator to review progress, discuss reclassifications and examine slippages and running expenditures. Several projects required the farming out of large chunks of design work in order to meet schedule milestones, or there were times when the NRC was consulted and project milestones were replanned and rescheduled. The nuclear managers at times were at odds relative to their areas of responsibility versus others areas of responsibility. Manpower loading of the work along with commitments within the nuclear organization was the only way to rationally commit to and perform the work. Managers are always shuffling individuals work assignments in order to meet organization commitments. That is the primary responsibility of a manager.

Life became more bearable after several reiterations of the LTP were accomplished. Work was being controlled in a logical, manageable fashion with minor perturbations along the way. Hindsight would have shown that the LTP should have been instituted several years before it actually became a dire necessity. One can only do so much work and spend so much money in a given time frame before something gives out. In this case it was utility managers who were "torched" in the name of safety.

If the LTP had not been instituted, one could foresee the regulators shutting down several utilities reactors until compliance was met. Post TMI and up to the present time, there has never been a time period when so much was demanded of the commercial nuclear industry in such a short time span. Today the regulator is back to normal everyday observations and inspections. Material modifications are practically non-existent with the exception of improving plant operating performance or for replacing major plant components due to wear and tear.

It is conservatively estimated that for those facilities designed and constructed in the 1970's and early 80's, the cost of safety improvements demanded by the regulator escalated the capital book value of these units four fold or more. A facility that was constructed for upwards of 200 million dollars had 600-800 million dollars of safety improvements without a single increase in plant electrical output.

Post TMI was a hectic time in looking back. Incredibly all the nukes did the regulator requested fixes and there has not been an incident in this country which would require these system perform their intended safety functions.

Chapter 17

The Nitty Gritty of Being on the Bad Guy List

Chapter 9 discussed getting on the Bad Guy List in general. The specifics of attaining such a dishonor evolved over a few years of marginal performance. Sometimes, in the heat of battle one forgets just where you are, are you winning or losing or merely striving to stay alive. The backlog of open items continued to grow despite big pushes to make it go away. When "goes into's" is greater than "goes out of's" the result is an increase in open items. The plant personnel could not overcome the existing day to day work load all the while attending to the backlog of items of concern, so they continued to rise. That became major problem #1.

The continued lack of progress in improving identified weakness in the NRC evaluated categories compounded the problem. It was realized that the open items backlog was one of the reasons the regulator refused to issue improving report results. INPO joined the bandwagon on the side of the enemy when its' open items list at the plant was not given elevated attention. In comical terms, you really don't give a hoot whose jersey the alligators are wearing while you are bailing out the swamp. An alligator is an alligator. INPO items, NRC items, QA/QC deficiencies, they all required manpower to resolve all the while the plant personnel were trying to keep the ship afloat.

The straw that broke the camels back came in the form of three operating abnormalities and they were not resolved in a timely fashion. This became major problem #2. One abnormally was the intersystem leakage in an emergency core cooling system. It was presumed to be either a valve leaking while shut or a heat exchanger suffering internal leakage. This phenomenon was recognized and plans were afoot to address it at the next plant shutdown. The second abnormality was that one of four main steam isolation valves, which isolate the reactor from the turbine, displayed deteriorating performance when operated and thirdly, the reactor master control switch, which is similar to the ignition switch in your car, had indications of erratic operation during startups and shutdowns. These three combined operating abnormalities raised the concern of the onsite resident NRC inspectors to the point where they contacted their superiors in King of Prussia, Pennsylvania. An inspection team arrived at the facility at the same time that the plant was being shutdown to address these problems as well as to perform a refueling outage. The regulator decided that now was as good a time as any to obtain the individual attention, not only of the onsite managers but the nuclear executives as well. They issued a "Show Cause Order" which in layman's terms says "your are shut down by decree of the NRC."

"You will locate, analyze, repair, test and closeout the three abnormal operating conditions. Additionally, you will identify, assign, implement and closeout your mountain of backlogged open items, and you will not restart the plant without first convincing the NRC that you have instituted management controls to preclude the recurrence of another mountainous open items backlog of work." The NRC literally took away the keys to startup the reactor. The plant proceeded to refuel the reactor, locate the source of the emergency core cooling intersystem leakage and repair it. Additionally, the plant personnel replaced the reactor master control switch which was a nightmare relative to ascertaining control of all the wiring which, in essence, interfaces with a multitude of control and safety systems. The mainsteam isolation valve problem was identified and repaired.

Senior executives from around the country were contracted to come and assist with planning and scheduling all the open items in the backlog. The first order of business was to assure ourselves that all the items, residing in several data bases were incorporated into the master database. The next order of business was to review each item in the context of whether or not it affected or was associated with other items. The items were scoped and assigned owners for implementation. Numerous items had major inputs on operating procedures and plant drawings.

Eventually, the backlog of open items was implemented and closed out with mounds of documentation generated. Now the NRC said, "Show us your management controls, your implementing procedures and your dedicated staffing to address future identified deficiencies." Then the next edict and request from the regulator was the crowning blow. The company executives were notified that the utility was required to submit a controlled, slow, plateaued startup plan. The regulator would install hold points(plateaus) to evaluate the companies competency to operate a facility in an approved manner. The hidden meaning behind those words were that "you are back to square one just like when we issued you your operating license. You must demonstrate that you have the hardware, the software and the personnel talent to operate a nuclear power plant satisfactorily and safely." The results of the shutdown and "show cause order" were that the senior plant personnel who operated the plant prior to shutdown were, in some cases, either relocated or replaced. New blood was hired to demonstrate the seriousness of the utility to operate with a renewed attitude toward successful performance. Many management control systems were modified or created to implement and actively address issues or deficiencies as they arose.

The final tally on this episode is that the plant was off-line for 33 months(3 months short of 3 years). The refueling outage and the subsequent expenditures of corporate funds totaled in excess of $335 million dollars.

What could not be performed to reduce the backlog of open items and successfully operate the plant simultaneously resulted in the expenditure of significant corporate resources all the while the plant was a non-producer of corporate revenue.

Sometimes it requires that you whack the mule across the nose with a two-by-four to get his attention! The NRC used a show cause order.

To those remaining, the burnout factor had just increased tenfold!

Chapter 18

The High Level Radioactive Waste Dilemma in the USA

There are primarily two levels of radioactive waste generated at a nuclear power plant, low level and high level waste. Low level radioactive waste can be in the form of solids or liquid. High level radioactive waste is primarily in the form of solids, like fuel or highly contaminated equipment. Solids suspended in liquids can also constitute high level radioactive waste. A future chapter will address a high level radioactive waste piece of equipment.

Low level waste usually consists of such things as rags, mops, paper, small pieces of equipment like hand tools, some spent resins from water purifying equipment, etc. Low level solid waste at the plant is processed in a utility building where it is packed into drums or large shipping containers and compacted. A low point drain on the shipping/burial container is used to ascertain that the moisture has been extracted via compaction. The shipping/burial container is sealed and shipped by flatbed tractor trailers to one of two site which accept low level waste, Hanford, Washington or Barnwell, South Carolina. Hanford has ceased accepting low level waste so all the low level waste is now sent exclusively to Barnwell. Once the container(s) arrive they are inspected for any transport damage and checked for degree of dryness by opening the low point drain to check for any liquid runoff. If there is liquid runoff, the container is not accepted for burial but returned to the generating facility.

Low level liquid waste is stored in storage tanks, allowed to decay, diluted and when satisfactory low levels of activity are obtained, the liquid is discharged overboard along with additional dilution. Let's look at the handling of high level waste. In the USA, there are no facilities for reconstituting spent fuel (used up fuel) as they used to be in France. The high level waste at US nuclear power plants are the reactor fuel which has been expended in the reactor and then, during refueling outages is transported under water from the reactor core to the spent fuel pool which resides besides the reactor. There the fuel, stored in fuel racks, decays slowly over eons of time until it becomes stable. During the reign of several presidents from Jimmy Carter to the G.W. Bush era, the US government was tasked with finding and developing a permanent storage facility for all the high level radioactive waste generated in this country. We have watched and waited for greater than thirty years for such a facility to be built. Presently there is not a permanent storage facility in service to accept the high level waste.

The requirements for the storage of high level waste are an absence of moisture over centuries of time and no geological perturbations for millenniums. Such a place resides in the mountains in Nevada. The place is called Yucca Mountain. This is a political football to say the least. Nevadans want nothing to do with the storage of high level waste in their state. It is called the NIMBY philosophy (not in my back yard). The governors of many states are adamant that high level waste shipped by trailer truck or rail is too dangerous to allow. The government will eventually shove this whole issue down everyone's throat and send the waste to Yucca Mountain. The facility has been under development for years but is still not ready for waste acceptance. Meanwhile the capacity to store spent fuel at the nukes in this country are rapidly decreasing. Something needs to be done, and as usual the government, when it cannot perform its functions, turns to the people and says we think it would be all right to store the spent fuel in dry storage facilities at each site. Now we will have more than 100 high level storage sites rather than one central site.

An article in the major newspaper dated July, 2004 stated that a federal appeals court threw the nuclear waste repository into doubt by demanding the federal government devise a new plan to protect the public against radiation releases beyond the next 10,000 years. An Energy Department spokesman said the agency was confident it could come up with a plan to protect the public against radiation for longer than 10,000 years. The government wants the Yucca Mountain, Nev. site as the repository while the anti-nukes want no part of it. Congress approved the site in 2002 over Nevada's objections.

This saga will continue through many more presidents, whereby the responsible Department of Energy secretaries say, "Let's pass it on to the next administration" or as they say in the trades, "not on my watch".

Chapter 19

The Radwaste Concentrator Fiasco

The role of fossil fuel minded corporate executives come to bear in this next episode. This occurred during the construction of the nuclear power plant and the time frame is around late 1971 or early 1972. As the plant was being constructed, the utility began staffing for the eventual takeover and operation of the power plant. Some in-house personnel were promoted and became the cadre for the eventual full staffing. Outside resumes were reviewed and many young up-and-coming individuals were hired along with many other in-house people to man the positions. When equipment was installed and the accompanying system was released from the constructor to the client, the instrument and control technicians were assigned to calibrate, pre-operational test and operate the system. One of the last systems to be turned over to the client was the radwaste system with its numerous and varied functions. All radioactive waste generated in the plant eventually finds its way to radwaste. There were numerous large storage tanks to hold the liquid waste, there was a compactor to solidify the low level waste and then there was the infamous Radwaste Concentrator. To begin with this huge boiler-compactor was a "rube goldberg" to say the least. Its construction was of poor quality, its instrumentation was questionable as to its performance and all-in-all the whole smear was a fiasco. The technician and mechanics who daily tried to calibrate it and test it returned to complain to their supervisors of the myriad of problems they encountered. Electrical EMT tubing was used for piping, the instruments would eventually be encased in insulation when it was applied, etc. Nothing worked as it should and the issues finally became so numerous that the manager of instrumentation wrote a multi-page memo highlighting the problems and prescribing solutions.

The purpose of the radwaste concentrator is exactly as the name implies. It would accept high level liquid waste, i.e. waste that contained solids which in turn caused the liquid combination to be highly radioactive. The concentrator would heat the liquid to extreme temperatures to boil off the liquid and leave the remaining solids as a sludge which could then be compacted. Once this equipment was placed in service it would reside behind a concrete and lead brick wall to shield it from emitting radiation to personnel in the area. The entire process was to be operated remotely from a control panel. Access to this equipment would be impractical as we will see later on. The memo was delivered to the companies Director of Construction whose job was to manage the onsite construction as an interface with the builder. He in turn reported to the Executive Vice President of the utility company. This nuclear power plant was the Exec. V.P.'s baby. He recommended it, he got the idea approved and he was going to see it through to operation, come hell or high water. Come to think of it, I guess it was the high water that sent the whole shebang to hell. Anyway, the memo was returned to the manager of instrumentation with the unwritten response from the Director of Construction that the Executive Veep wants this mother operational as soon as possible, to implement your recommendations would add months to the construction phase and delay startup. In hindsight, no one fought hard enough to see that the plant could in essence be started up and the concentrator modifications/improvements made long before it was necessary to run the concentrator. But politics being what they are, the concentrator was tested and accepted with its many foibles.

During plant operation the time came to process high level radioactive liquid waste and the system was started up. It was noted that it performed for a short period of operation before unacceptable results were received. Liquid waste was exiting the concentrator and contaminating the equipment and the floor. The sludge transfer was unacceptable and the use of the concentrator was terminated. Several years later an evaluation was performed to see if the entire concentrator could be dismantled and shipped off site for burial. Attempts were made to enter the sealed off area to determine what the radiation exposures would be to individuals who would prepare the area and equipment for dismantling by remotely operated equipment. It was found that depending where in the area the personnel operated, the stay time was from four to fourteen minutes. That means an individual who went into the area received his quota of radiation which would preclude him/her working in any radiation environment for the next three months. They would receive their three months allotment of radiation, under federal guidelines, in four to fourteen minutes. The entire project was scrapped and the concentrator room catacombed, as it probably remains today.

The shortsightedness of an executive with a fossil fuel mentality caused this whole fiasco. He did not investigate alternatives, he did not question the issues to the void and he did not ask for recommendations. He placed his sights on getting the plant on line as soon as possible. The radwaste concentrator tomb should bear his name for posterity.

Chapter 20

Main Steamline Isolation Valve Inflatable Test Plugs—Yea or Nay?

This chapter deals with whether to buy test plugs for testing valves during an outage. The subtle undertone of this chapter is that a major impediment to the daily operation of this nuclear power plant was the interference, the incompetence and sometimes the outright arrogance of executive management and their decisions.

Follow along. During the planning stage of an upcoming refueling and maintenance outage, a systems engineer, who had the responsibility to test and document the integrity of all isolation valves that communicate directly with the reactor vessel, suggested the company purchase inflatable test plugs. The description of the valves function means that all valves which are required to isolate the reactor vessel (bottle it up) are required, by code, to be tested on a regular frequency and their leakage must not exceed code requirements. The way to test these valves is to close them, defeat any controls which could inadvertently open them and pump either air or water against the valve plug to a specified pressure, then observe the amount of leakage (pressure drop) between the valve plug and valve seat. Excessive leakage requires the valve be dismantled, inspected, repaired or replace internals and retest until acceptable test results are obtained.

The main steamline isolation valves are huge valves which shut when required to do so and stop the flow of high pressure steam from exiting the reactor and flowing to the steam turbine. There are two sets of valves in each steamline. One valve is in the primary containment building close to the reactor vessel and the second valve on each steamline is outside the primary containment building, in the reactor building (redundancy). The systems engineer who suggested the purchase, by the company, of the four main steamline inflatable plugs, foresaw the expediency of installing the plugs and inflating them such that this would reduce the time required to test the valves and also assist in identifying which, either or both, of the valves is leaking. The way we tested for leakage without the plugs was to close both valves on each steamline and pressurize between the valves. One valve was being tested in the steam flow direction and the other was being tested in reverse of the steam flow direction. The price of the plugs was greater than $25,000. Had the price been less, the request for the purchase of the plugs would not require executive nor Board of Director approval but could be authorized by the Nuclear Vice President.

The Refueling Outage Manager, in concert with the Engineering Manager, the Operations Support Manager and the Plant Manager all approved the benefits to be derived and therefore the purchase of the plugs. The Operations Support Manager prepared the capital authorization request for management/corporate approval. The plugs were available for immediate delivery from the nuclear steam supply designer/vendor of the facility. Approvals were obtained and the Vice President-Engineering asked the Operations Support Manager to make the presentation to the Senior Vice President (fossil and nuclear) as the proponent of the authorization. Approval by the Senior Veep would then result in the authorization being placed on the agenda of the next meeting of the Board of Directors.

The Senior Veep was inquisitive to a point but not totally convinced the plugs were a "good buy". He heard the positive aspects of the issue but reflected on several negatives such as: "We tested these valves on several previous refueling outages without these plugs"; "Not all facilities use these plugs"; "Why should I spend funds for something I don't think we need?"; "You guys down in nuclear are always inventing ways to spend the companies money." Notice he did not say " Show me how we are saving money by expediting the return to service."

"Where is the five year payback?"

Instead he exerted his executive prerogative and vetoed the purchase. The Operations Support Manager persisted in trying to convince the Sr.Veep he was making a foolish mistake (wrong choice of words this morning) and his decision was not in the best interest of the company (second boo-boo). Those words were the final discussion. He dismissed the manager with the edict that he was, for all practical purposes, on this issue, "the company" and his decision was final. His shortsightedness, his ego, his swagger were counterproductive to the success of the refueling outage but he never really understood that. The manager had finally met someone who epitomized the Peter Principle, you are promoted upward until you reach your level of incompetency. Some heated words were exchanged reflecting upon superior/subordinate relationship and as the manager was exiting the Senior Veep's office the Veep said something like "One of my last great acts around here will be to fire you." Whereupon the manager turned and in a pissed-off manner said "And where are they going to find another asshole to replace you?"

No sooner had the Ops. Support Manager returned to his office, convinced that the Senior Veep must be undergoing senility, when his boss, the Nuclear Veep, asked for his presence in the Veep's office. The conversation was very civil. The Veep knew the importance of these plugs. There were no admonishments, no recommendations, no "let's try again".

This story has a good ending. The Ops. Support Manager, with the concurrence of the other managers who understood the value of purchasing the plugs, agreed to pursue the "acquierment" of the plugs.

A representative of the nuclear steam supply designer/vendor was called and told to reserve the plugs for use at our facility. The plugs were delivered to the site without an approved, plug specific, purchase order. Eventually, the purchase of the plugs was 'buried" under something like "expendable tools."

Sometimes middle management, more cognizant of the true needs of the facility, had to go and take the bull by the horns regardless of the consequences. This is another nail in the coffin of fossil fuel minded corporate executives who didn't know "their ass from a steamline plug." It is believed that the Senior Veep went to his grave believing he exerted the proper executive skills to his decision. So be it!

Chapter 21

Material False Statements or You Better Tell It Like It Is

The experience of working in the commercial nuclear power industry was not already overbearing enough in its requirement to toe the line, when along comes a new law entitled "material false statements." It doesn't matter whether you make these statements under oath and affirmation or not. If you make a statement to the regulator which is false, either orally or written and is of a material nature, it is a material false statement and carries varying degrees of penalty with it. What is a material false statement? It is an utterance either spoken or written whereby the content of the statement leads the regulator to believe it as fact. Later on the essence of what was presented as fact is deemed to be not as stated. For example, you correspond with the regulator and sign you name to a letter from your employer attesting to a condition or status which later on proves to be not entirely factual.

This problem happened and was widely circulated throughout the nuclear industry. A midlevel manager at a commercial nuclear power plant submitted correspondence to the regulator stating that, as committed to in a previously submitted schedule, all the emergency preparedness speakers were installed to alert the general public in the event of a nuclear mishap. This midlevel manager had been told by his subordinates that they indeed had been installed. Now this manager did not take it upon himself to go out into the countryside and personally verify that what had been told to him was indeed fact. He took the word of his fellow employees that it was as stated. It so happened that a speaker was found to be defective and thereby not installed. His correspondence to the regulator constituted, in the eyes of the regulator, a material false statement. Said manager eventually was relieved of his job and assigned elsewhere. Material false statements laws were intended to allow the regulator to accept, as fact, that which was presented to it.

Utilities ultimately were required to implement internal procedures to preclude the utterance of material false statement. This utility required several levels of assurance before it submitted any statement of fact to the regulator. Letters, both those requiring signatures under oath and affirmation and those not requiring affirmation, were circulated around the utility and any entity who had any input to the written word was required to sign off as to their knowledge and concurrence as to what was stipulated as fact. Letters requiring signature under oath and affirmation were signed solely by corporate executives and notarized. One can only imagine how strenuous this process was along with all the other requirements and conditions under which one had to toil.

Several managers were interviewed by personnel from the regulators Office of Investigation and Enforcement as to whether they had made material false statements to the regulator. Degree of misstatement or the degree of the implications of the response were not significant. Materially false statements were all categorized together. Small or big in implication mattered not. There were days when it was tough to go to work. Envision what would happen to our honorable elected representatives if they were under the same microscope. How many politicians would be in jail? How many executives have you heard say they had no prior knowledge of what they are being accused of? How many executives have denied openly a fact that was later revealed to have been so? (Remember the cigarette executives before congress?)

The nuclear power industry was born as a peaceful use of the atom after the military use of the atom as weapons of mass destruction and realize that a nuclear mishap can have terrible consequences to the general public and the earth itself but the regulator, in their quest for stringent application of the law, sometimes forgot that it is acceptable to bend but not break.

The commercial airline industry would not have one single airplane to fly if the FAA imposed the exact same criteria on the airline industry as it imposed on nuclear power. You have heard the industry say that a FAA recommended fix would be to costly to implement. It's all a matter of degree. An airline crashes and 300 plus people are killed. A nuclear accident happens and thousands of people could be killed!

The author would be elated if the government imposed material false statement requirements on every industry and every executive in this country. The application of material false statements to the correspondence between the utility and the regulator only added to the burden borne by those who worked in the commercial nuclear power field. It was another escalation in the requirements to play the nuclear power game. Burnout was caused primarily by the myriad of requirements that were imposed to remain in compliance.

Some people never even realized they were in advanced stages of burnout despite the fact the flames were licking at their asses!

Chapter 22

Do as I Say, Not as I Do—the NRC Credo

In the foreword, the author mentioned that the regulator, particularly the NCR is both the issuer of the license to operate and the regulator, something like your Registry of Motor Vehicles. You go to the Registry, take a driving test, written and on the road, and get your license to operate a motor vehicle. There are, at times, Registry cops out patrolling the roads and they can pull you over and give you a ticket or even rescind your license. They are dual purpose entities unto themselves. So to is the NRC. It has been tasked by Congress to oversee the nuclear power industry in this country but it also oversees, regulates and monitors the government facilities which fabricate nuclear materials for the defense of the country.

Again, we have all read in the paper or seen on television where reporters have interviewed workers at some nuclear processing plants such as Oak Ridge in Tennessee and the facilities in Colorado, Ohio and elsewhere, where the exposure to radiation by the workers or the despoiling of the earth and water are outside the letter of the law. None have been fined, none have been shut down nor have any been openly reported as to the degree of severity of the offenses committed. There is a dual standard in play. The government facilities, for the most part, are contracted out to large corporations such as the Westinghouses, to operate them for the government. Radiation exposure to anybody, regardless of who employs them, is radiation exposure, period. The radiation does the same degree of harm to the skin and body organs whether is comes from a manufacturing facility or a nuclear power plant or a research laboratory. But the government facilities, if they do provide exposure reports, are not for public scrutiny. Prying reporters only find out the goings on at these facilities by poking around and finding disgruntled or injured workers willing to speak out. Why do we have dual standards? Why aren't the routine reports mandated by the regulator of public utilities also mandated of the government facilities? The "do as I say, not as I do" ploy is in play. When have you ever read in the paper where a contractor operating a government facility has been fined thousands of dollars for not correcting serious deficiencies at the facility? Never to my recollection. Aren't the serious infractions there as serious as those at a power plant? Wouldn't the catastrophic results of a serious mishap at a government facility be as injurious to the general public as one at a nuke plant? I guess the question one should ask is "Who regulates the regulator?"

I guess the answer, depending upon who you ask, is either "the regulator or nobody".

Having toiled under the laws that regulate nuclear power, one gets upset when they remember all the pressure imposed by the regulator on the utility employees. Every thing must be as they request. Nothing is good enough. Omissions of insignificance are unacceptable. Yet one wonders where these same feds are when the government facilities perform at less than legally accepted standards. The blanket of national security is to often spread too broadly to envelope unacceptable practices and unacceptable results at the government owned facilities. Are not the general public who reside near these facilities entitled to the same protection as the general public who reside near a utility owned nuclear power plant?

I guess this all flows down from the federal government itself who have their own rules and their own benefits which far exceed those of John and Jane Doe.

Do as I say is all well and good only if it is followed by do as I do!

Chapter 23

A Month in the Life of a Nuclear Power Plant Manager

Let's walk in the shoes of and sit in the chair of a nuclear power plant manager for a week or more. The following chronology is typical, not exact, for purposes of trying to capture the scope, breadth and pace of this middle-management manager position.

Monday morning—7 am—"The morning meeting" where the various disciplines present the status of work that was to be performed in the past twenty-four hours and what is on the daily schedule for completion or implementation in the next twenty-four hours.(Notice this meeting does not cover what was done over the weekend. Those morning meetings took place each morning during the weekend. Electrical power production is a twenty-four hour per day, seven day per week job. Weekends are just two more day in the week.) This would include major and/or routine maintenance, operational surveillance testing, equipment calibration, radwaste shipments and off-loading of liquid waste, any new QA/QC deficiencies and what will be performed to overcome any schedule slippage. Engineering will input on status of equipment purchases. In attendance may be an NRC resident inspector. The meeting is usually chaired by the plant manager if the plant is on-line or the assistant plant manager on weekends. If the unit is shutdown, the outage manager chairs the meeting. During shutdowns, where the plant is worked twenty-four hours a day to return to service, two shifts of twelve hours each are worked and there are two morning meetings, usually at 6 am and 6 pm.

Next the plant manager usually calls the Nuke Veeps with update reports relative to issues brought up at the morning meeting. Following that the plant manager usually would meet with the assistant plant manager to discuss such issues as status of the expense or operating and maintenance budget, any snags with resolving QA/QC deficiencies, backlog of work status, etc.

The resident NRC inspector may request a meeting with the plant manager to obtain information relative to an issue of concern to the inspector. There are revised procedures(many) for his signature. The refueling outage manager may have an issue he/she needs clarification on with respect to plant conditions required, before he/she authorizes the work to be included in the outage scope.

The plant managers day primarily consists of transfer of information so he/she can receive and dispense information for the successful operation of the entire facility. Tuesday through friday consist of the same routine as previously described, with the addition of a least one or more important meetings either with the NRC Office of Inspection and Enforcement, the QA/QC Manager, the Engineering Manager, INPO, insurance company agents, etc. The NRC inspects and audits every aspect of plant operations. Therefore it has inspectors for every aspect of that performance. For example, it would inspect maintenance performance including electrical, mechanical and instrumentation and control maintenance. It would monitor radioactive waste performance including liquid, solid and gaseous releases. When it says monitor, inspect, audit, that refers to documentation, tons of it. The NRC looks at documentation of what was performed. They also would monitor plant chemistry which primarily deals with the quality and purity of water that interfaces with nuclear steam generation. It monitors and inspects security, engineering, emergency preparedness, training, modification/construction and configuration control which would extend over into documentation and records management. They inspect and monitor performance for fire fighting qualifications and training. I'm sure I have omitted several other disciplines/operations that are inspected. The idea here is to present the reader with the multitude, the vastness and the scope of work that the plant manager is involved with. These inspections are staggered and may occur several times per year or only once per year depending on the degree of concern in the mind of the regulator. These inspections are independent inspections and two or more may be going on simultaneously. When NRC Inspection and Enforcement inspectors appear at the gate, either announced or unannounced, they commence with an entry meeting to define what the inspectors are here for and what they will be looking at. The inspections also conclude with an exit meeting where past open items are discussed, new deficiencies or areas of concern are brought up and discussed in length. These entrance and exit meetings generally take place in the Plant Managers office. The NRC Resident inspectors hold a monthly meeting with the Plant Manager likewise.

INPO holds entrance, exit and sometimes end of day meetings if they (INPO) have serious findings which they want brought to the fore immediately for consideration and possibly resolution before the exit meeting. All inspections by the various regulators result in written reports. The findings which require attention are placed as work items in the master work control database and all require written responses for future items of inspection before closeout by the regulator. Guess what? The backlog of work just got bigger!

Plant discipline managers may request a meeting with the Plant Manager if they have a concern which requires elevation or at least information for the Plant Manager, so he won't be blindsided.

The Plant Manager may appear to be King Solomon who knows the right answers to all problems. Sad to say, the Plant Manager is like an orchestra conductor, he needs to lead in order to make sweet music. An old adage in management is to "avoid leaping monkeys". Someone comes to his office and presents a problem. The Plant Manager says the magic word "I" will look into it and get back to you. The person with the problem just had a leaping monkey jump from his back to the Plant Managers back. He now owns the monkey!

The Plant Manager's month goes along just as previously described with one or more exceptions. If the plant "scrams" (trips off-line) or shuts down in a controlled manner to investigate an abnormality of plant operating parameters, the burden on the Plant Manager just doubled. He must direct the plant personnel to evaluate and prescribe corrective actions to reduce outage time and get the plant back in-line. Additionally, the Refueling Outage Manager is constantly interfacing with the Plant Manager about scheduling routine and/or major maintenance, refueling and major modification implementation to produce the shortest outage duration schedule.

Okay, there are scads of meetings, info coming in, info going out, problems to address, decisions to make, budgets to control, radiation exposure to plant personnel to reduce to "As Low As Reasonably Achievable" (ALARA). There is the monitoring of the backlog work and its growth. No wonder, we just had several inspection findings reports and all unresolved open items are inserted into the master workload program which includes the backlog. There is the routine management tasks such as: plant personnel performance evaluations, raises to be distributed, annual budgets and manpower requirements to prepare, present, get approved and track. There are new personnel to interview and hire and non-performers to escort out. There are union grievances, turf wars to settle, egos to stroke. One can understand why some jobs make you bounce off the wall to avoid material false statements and the need to juggle work schedules to allow licensed personnel to receive mandatory training to guarantee nuclear safety and configuration control. There are fire drills to conduct on prescribed schedules including the fighting of electrical fires with water. There are emergency preparedness drills to conduct, monitored by the NRC and FEMA (Federal Emergency Management Agency) to ascertain the competence to identify plant mishaps, calculate the release of radiation to the general public, alert the populace and confer with and make recommendations to the governor of the state.

There are phone calls to make to the NRC in the event of abnormal or unusual plant operations. There is INPO to notify if safety system related components fail so INPO can insert the data into the Nuclear Plant Reliability Data System (NPRDS).

The Plant Manager has a twenty-four hour per day, seven days per week job. His pager goes off frequently when he is not on-site.

Is it any wonder that the job is practically impossible to be performed by one individual or even two individuals? There is no room for hesitation, reflection, delay or error. There are never enough hours in the day to make all the decisions expeditiously. The job is literally programmed to fail. Failure begets burnout and burnout begets more failure. Most plant personnel work forty hours plus varying degrees of excessive overtime. The plant manager works a couple of weeks per week! There are other jobs, previously mentioned, which likewise incur burnout. People work long hours to perform inordinate amounts of work to reduce the backlog of work, only to find burnout beats completion. Personnel are not replaced because of burnout, they are replaced for failure to perform. Prophetically, they failed while over-performing. Only in the nuclear power plant business can such lofty achievements lead to personal failure. One remembers the old expression, "it's like shoveling shit against the tide". Someone once said, "Boy, being in the nuclear power plant business must be fantastic!" If they only knew.

Chapter 24

The Eighty Percent Comfort Zone—Outage Management

If it was not previously mentioned, the high burnout positions included the Plant Manager, the Outage Manager, the Vice President of Operations, Maintenance Division Manager, the Training Department Manager and the QA/QC Manager primarily. All are continuously under the nuclear microscope of the regulator and utility executive management. One of these positions developed into a full time job, that of the Outage Manager. Originally the Plant Manager was responsible for the day-to-day operation whether the facility was generating power or shut down for maintenance, refueling or major modification installations. This proved to be the killer of all killers. He/she was responsible for managing everything while the plant was operating and in addition, to plan, schedule and implement the outage schedule when the plant was shut down.

Senior management eventually saw the light and designated an Outage Manager. This year-round job required the integration of all the work required to be accomplished when the plant was shut down. The planning and scheduling is, and will continue to be a Herculean task. Everybody wanted all their areas of responsibility work to be scheduled while at the same time had not the least idea of what that work entailed. They could define the scope of the work and they could tell you that it can only be completed while the plant is shut down. The term "like pulling hens teeth" adequately defines the degree of difficulty that existed in trying to extract all the pertinent information to allow the job(s) to be planned and scheduled and also integrated into the master schedule. "How many men will the job require?" "What trades?" "When will the final design be completed?" "When will the critical materials arrive onsite?" These were but a few of the questions requiring answers so as to effectively schedule the work. The goal of the Outage Manager is to complete all the work required to be performed in the shortest time frame. This shortest time is known as "the critical path." A facility that is shut down is not only a non-contributor to corporate income but in fact is a corporate expense. Instead of generating power, it is consuming power.

After several outages, some successful and some not so successful relative to downtime, it became apparent that the Outage Manager was always the fall guy when the outage was not declared a success and was only a player in the outage which was completed on-time and at or under budget. Thus the term "the 80% comfort zone" came into being. It meant, take over the outage management job after all the work has been planned, scheduled, manpower loaded and ready to roll (the first 10%). Also, get out of this job before the last 10% is completed. One could not be held responsible if the plan and schedule were faulty and one could not be held responsible if the job ran over the schedule or costs soared. Also the 80% manager need not face the DPU after the plant went back on line to explain to the regulator how and why the outage was implemented prudently and therefore the company should recover outage expenses from the consumer. This last task of the outage managers job was a "sleep loser" to say the least.

The successful 80% Outage Manager escaped the fires of burnout and the responsibility to plan, schedule, and implement the next outage. Ode to joy! An Outage Manager's job life span was very short to say the least.

Chapter 25

A Layman's Description of the Three Mile Island Accident

March 28, 1979

The accident at Three Mile Island in Pennsylvania caused the commercial nuclear power industry's expansion to come to a screeching halt. What was once described as Americas way to significantly reduce its dependence on foreign oil came to a dead stop after TMI. Utilities with partially completed nuclear facilities moth-balled some. Others canceled orders for new units. There were 109 operating nuclear power plants in the USA and there have not been any orders for new plants post TMI.

In layman's terms, what happened at TMI, why and what were the consequences at the facility itself follow. Let me begin by stating that there were no fatalities at TMI and no one was over-exposed to radiation at TMI. The plant was operating normally. It is of Pressurized Water Design (PWR) which in layman's terms means it has two separated water loops to produce steam. Water is pumped through the reactor and nucleate boiling develops. Nucleate boiling is similar to the bubbles that form at the bottom of a pot of hot water just before it starts to become a rolling boil. The steam is sent to one side of a steam generator, another word for a heat exchanger. Water flows through the other side of a steam generator and the steam gives up its heat to the water through the sides of the tubes making a secondary steam loop. The reactor steam and the secondary steam never come into physical contact with each other. A PWR has a pressurizer tank on top of the reactor vessel where the steam collects to pressures as high as 2500 lbs. per square inch. The pressurizer at TMI, like all high pressure vessels, has safety/relief valves per code requirements. They are used to relieve pressure when called upon to do so. At TMI a pressure relief valve was operated and then failed to close. Steam was continually relieved from the pressurizer which in turn caused more steam to leave the reactor vessel. An increase in steam flow out without an equivalent water makeup results in a reduction in water inventory in the reactor vessel where the nuclear fuel resides. Water flow through the reactor vessel serves three purposes. One, it cools the reactor fuel by keeping a constant inventory of water above the top of the fuel. Two, it moderates; that is, it serves to slow down the speed of escaping electrons such that they can be captured by surrounding fuel to make more nuclear collisions which produce heat to generate steam. Reactor physics is based upon electrons being slowed down to be captured to cause more electrons to escape, etc. Three, it shields the surrounding areas from the core's radiation.

The basic premise in operating commercial nuclear water reactors is to always keep the reactor fuel core covered with water to prevent overheating and fuel meltdown. The TMI operators, observing several parameters which in their minds were contrary, stopped the insertion of flooding water to the reactor core. In short, the core suffered a lack of adequate water inventory to maintain the fuel covered. The heat buildup was such that a portion of the reactor fuel overheated and slumped into a molten mass on top of the rest of the fuel core. In hindsight, had the operators sat down with their hands in their pockets, the emergency core cooling safety systems would have operated as designed and prevented the mishap. Human intervention in the wrong direction was the culprit.

Multi-millions of dollars were subsequently expended by the US commercial nuclear industry to provide redundancy of equipment to analyze reactor parameters, to provide systems like post accident sampling which are capable of obtaining, remotely, gas and liquid samples for analysis as to what is happening or did happen in the reactor and surrounding regions of the core. Emergency Preparedness grew to a multi-headed entity with literally hundreds of activities to prepare for and implement, if required, the protection and/or evacuation of the general public. The TMI ramifications on emergency preparedness would fill several books. The TMI mishap was the single event which drove a stake through the heart of the nuclear power industry. Never has so much effort been expended in human and monetary cost to preclude another TMI but also to recover if such an event should occur. Twenty plus years after TMI, the nuclear industry has still not recovered sufficiently to resume the expansion of commercial nuclear power.

TMI was the single event which put the regulator into a frenzy of demands of the industry to accomplish in an environment non-conducive to practical work efforts, in my opinion. Everything was an overbearing surge to put the industry into a posture to preclude and/or overcome another TMI.

The following excerpts are gleaned from the Kemeny Commission Report, formally known as "Report of the President's Commission on the Accident at Three Mile Island".

The accident at Three Mile Island Nuclear Power plant occurred on March 28, 1979. President Carter appointed John Kemeny, President of Dartmouth College to head up the investigation of the accident. The Commission presented its report to the President on October 28, 1979.

I will list several pertinent findings of the commissions report but they are too numerous and voluminous to cover in detail or specificity.

-There was a belief that the nuclear power plants were sufficiently safe. This became a conviction. This attitude must be changed.

-The NRC, responsible for the safety of nuclear power plants, imposed regulations upon regulations such that safety became a negative factor in nuclear safety. Too much emphasis was placed upon safety of equipment with the downplaying of the importance of the human element in nuclear power generation.

-There were serious inadequacies with both the licensing function and inspection and enforcement activities at the NRC. The commission concluded that there was no well-thought-out, integrated system for the assurance of nuclear safety within the current NRC. The present organization, staff and attitudes at the NRC is unable to fulfill its responsibility for providing an acceptable level of safety for nuclear power plants.

-Human error was to blame for the degree of physical damage to the facility.

-Operator training for the adequate operation of a nuclear facility is greatly deficient.

After reviewing, again, the Kemeny Commission Report and in hindsight, the author believes that many of the recommendations directed at the NRC were not implemented. The multitude of recommendations for facility upgrades, operator training, emergency preparedness, etc. were implemented at all the nuclear facilities.

Note: The prologue of the Kemeny Commission Report provides a succinct account to the accident along with plant diagrams.

Chapter 26

Chernobyl—the USSR's Experiment with Disaster

Before one can explain what happened at Chernobyl to cause the most disastrous nuclear accident in the history of nuclear power, one must understand several basic premises about power generation in general. There are several well understood operating characteristics which are used to generate electricity via steam power. In a conventional power plant which generates electricity by the burning of fossil fuel (coal, oil, gas) in a furnace with a water boiler incorporated to generate steam, the terms used to describe the interface between the boiler and the turbine generator are these: the generator/turbine is the master and the boiler is the slave. Let's examine this in the same context as you driving your car. You are going along at a constant speed and you come to a rise in the road. If you do nothing, the car will decelerate as it climbs the rise, so you put your foot on the gas pedal and press down on it. The car responds by the carburetor opening its ports wider, more fuel is consumed and the car accelerates up the rise faster than had you done nothing. The generator (you) saw the need for more power because of the rise in the road (the load) and asked the turbine (the car) to go faster and the carburetor and the engine responded (the boiler) to satisfy the demand for more power. The same applies to a fossil fuel plant. Central dispatchers watch the generation of power versus the consumption of power and when demand starts to increase the dispatcher calls generating stations (fossil) to generate more power. Some plants have excess reserve capacity, i.e. they are not running at full power and can put out more megawatts. Others may be called upon to start up, etc. So the reduction in voltage out on the grid because of excessive power demands requires more power to bring the system back in balance. Remember, the boiler "hops to" and puts out more steam by burning more fuel to satisfy the generators demand.

Not so in a nuclear power generating plant. The reactor is the master and the turbine/generator is the slave. It was once said that a nuclear reactor is like a bucket of wet cement. You have to kick it to get it to move. Anyway, a reactor generates more power by either of two means. First it can increase the water flow through the reactor to sweep away more nucleate boiling bubbles and generate more steam. If this is not available then the reactor operator must withdraw control rods in the reactor core so as to expose more fuel surface area to surrounding fuel, therefore causing more reactions within the fuel and generating more heat to make more steam bubbles. These two activities are not allowed to operate automatically, but are done by human intervention under controlled circumstances. Therefore, regardless of what the turbine/generator demands, they are slave to the reactor. Power is increased in the reactor slowly therefore electrical power increase is "the tail on the dog" and the dog wags the tail, not vice versa.

That understood, let's examine what happened at Chernobyl. The reactors in the USSR were not water moderated but the fuel resides in a carbon pile. How this works is not relevant for this explanation. For now the discussion will center on what the human intervention or lack thereof did at Chernobyl. Contrary to nuclear physics and power theory of operation, the operators at Chernobyl were messing with the master/slave theory of operation. The operators decided to do an experiment to see what would happen if the turbine/generator were the master and the reactor was the slave. They systematically decided to request more power by making the turbine master and not do anything to the reactor to compensate for more demand. They squeezed all the power they could from the reactor and consequently the reactor overheated, melted and because the containment building(s) for Russian reactors are not as sophisticated as US reactors, excessive amounts of high level radiation escaped into the surrounding areas and the atmosphere.

This is a layman's definition of what happened at Chernobyl, with sufficient description to allow the reader to grasp the issues. Helicopters were required to drop cement into the containment to encase the reactor so as to slow down the radiation releases. Many of those helicopter pilots later died as a result of receiving excessive radiation exposure.

The Russian operators literally took it upon themselves to mess with the bull and not only they but many individuals got the horns. Chernobyl's destroyed reactor in now a tomb! If the US nuclear power industry wasn't dead, Chernobyl drove the finishing stake through its heart!

US reactor operations precludes this type of operation. Technical specifications (the license) describes how to operate the reactor. Violation of these procedures would constitute a flagrant violation of federal law and would be either sabotage or willful negligence, subject to stiff penalties.

Remember, a nuclear reactor is not placed in automatic operation at the plant. A load dispatcher cannot externally request that the reactor put out more steam to meet load demands. He can call the nuclear facility and request that, if the plant is not operating at 100% power and is capable of doing so, that the operators increase reactor steam flow so as to generate more electricity. It requires operator manual intervention to increase or decrease power. A nuclear power plant is not operated as a load follower unit as are fossil fuel plants. Nuclear power plants are operated at maximum capacity and the fossil fuel plants are used to increase or decrease output to meet load demands.

Chapter 27

So What Protects Us from the Radiation of a Nuclear Power Plant?

This question has been asked by the general public on numerous occasions and I am not sure a detailed explanation was provided. Several times the term "defense-in-depth" was mentioned without detailed descriptions of what defense-in-depth means.

Let's examine what exists between you, the general public, and the nuclear fuel(the source of radiation) in a nuclear power plant. Again, in layman's terms it goes like this. Nuclear fuel consists of enriched uranium. Uranium as it exists in nature tends to be unstable, that is it tries to realign itself to a lower element, physically speaking. Uranium ore is mined, crushed, slurried, enriched to about 2% in a centrifuge and then shaped and constructed into pellets. These pellets are about the diameter of a cigarette and are encased in ceramic (shield #1). The ceramic pellets are placed in hollow tubes called fuel rods about twelve feet long and capped (shield #2). Forty-nine twelve-foot fuel rods, placed in a seven by seven array, are strapped together to form one fuel bundle. This fuel bundle is surrounded by a twelve-foot plus square channel and a lifting handle is attached to the top to allow for transport underwater (shield #3). When four fuel bundles are inserted in the reactor, a control rod which is a long blade in the shape of a cross is inserted between the four fuel bundles and controls the activity of that fuel array. The control rod is shield #4. Water surrounds the fuel in the reactor core(shield #5). The entire core with its associated control rods reside in the reactor vessel(the pressure cooker). This is shield #6.

The reactor vessel is housed in a primary containment building which consists of thick steel, in the shape of an inverted light bulb, surrounded by six feet of high density concrete(shield #7). The primary containment resides in the reactor building. This is shield #8. There are eight shields to protect the general public from the reactor core fuels radiation. This is called defense-in-depth. Feel better now about nuclear power?

Chapter 28

The Browns Ferry Cable Spreading Room Fire and Its Aftermath

In the mid 1970's an accident occurred at the Browns Ferry Nuclear Power Plant in Alabama. Several times in previous chapters mention was made of this fire and its' ramifications upon the nuclear power industry.

Using layman's terms again, let's describe where the fire occurred, why it occurred, and what the consequences were to that plant itself. If memory serves me right, the plant experienced some significant difficulty in shutting down the plant and placing it in a cold condition, i.e. the reactor is shut down and no residual heat is observed.

In a nuclear power plant the control room is the nerve center of the facility. All important activities are controlled from the control room. Pumps are started or stopped, electrical breakers are opened or closed, fans are started, transformers energized, the reactor is started or stopped. Increases in power output are controlled from this nerve center. Indicators(lights) and alarms inform operators of what is running, where there may be trouble, etc.

Because it is the nerve center, thousands of cables are required to bring information to the control room. Cables are required to send signals to system equipment to start, run or stop. To handle this multitude of cables, a concrete walled, airtight room called the cable spreading room is constructed below the control room. All cables going to or coming from the control room pass through the cable spreading room (aptly named). Penetrations where cables pass through walls are sealed as are penetrations in the ceiling of the cable spreading room. Only when all penetrations are sealed is the cable spreading room airtight.

Now we know where the fire occurred relative to the control room. How did the fire get started in the cable spreading room? Construction crews, namely electricians, had installed additional cables either through existing penetrations or possibly through new penetrations in a wall of the cable spreading room. Their next job was to seal the penetrations to make the room once again airtight. They applied expandable foam which filled the penetration and sealed around the cables. Having done this, the electricians now gave the penetrations the "old fashion" leak detection test. They lit candles and placed the flame near the just sealed penetration. If the seal was not airtight, the flame would waft in the direction of the air flow, either into the room or out of the room depending on the air pressure on either side of the wall. In applying the old fashion leak detection test, some of the rubber jacket of the cables heated, ignited and spread. Before the electricians could extinguish the fire, the room had filled with noxious black rubber generated smoke.

Operators above the cable spreading room, recognizing that there was a fire below, attempted to shut the plant down(the most secure condition for a reactor when in doubt). They encountered difficulty in completely operating from the control room. Luckily, procedures were in place to allow the operators to go to remote locations and shut down equipment locally. Breakers were tripped or energized locally to stop or start required equipment. Eventually the plant was shut down and placed in a cold condition.

The NRC investigated the occurrence and concluded that modifications would be required to more expeditiously control the plant should the control room not be available(i.e. what if the fire was in the control room itself?)

Appendix R to the Code of Federal Regulations for nuclear power plants was enacted. This mandated that nuclear power plants were required to re-analyze the existing construction at each facility. A fire was postulated anywhere in the plant and the analyses required documented demonstration that the plant could be shut down and cooled with the entire loss of cable in the postulated area. Needless to say, the analyses concluded two results. The first result was that if you needed the cables in the fire area, they would require protection from the fire, i.e. be encased, or second, if encasement didn't prove successful in protecting the cables from the generated fire then either reroute the cables to another location or build a new location.

The first solution was that cables would be encased, after testing laboratories conducted numerous fire retardant applications, to protect the cables from fire. Some areas of the plant that contained minimum fire loading (flammable materials) were allowed to encase the cables in approved fire retardant materials. Other areas of the plant required relocation of critical cables outside the heavily fire loaded areas.

Appendix R modifications continued on for up to twenty years after the Browns Ferry fire. In some cases, cable duct banks were constructed on the outside perimeter of the facility. That was the only way to have cables pass from one area of the plant to another area, by putting them in conduits which were then encased in cement and buried outside the plant. Assurances were highly probable that that was the most conservative approach to establishing compliance with Appendix R of the Code of Federal Regulations.

Plant shutdown capability from outside the control room was also greatly enhanced post the Browns Ferry fire.

The fire accident at Browns Ferry was but another episode where a sister nuclear utility experienced a major problem and the entire nuclear industry was required to analyze and determine if their facility could overcome a cable spreading room fire and safely shutdown. That scope was expanded when the analyses required safe shutdown with the loss of any one area of the plant due to a catastrophic fire. Like TMI, multimillion dollar modifications were required to comply to a new found trouble.

Footnote: In areas where light fire loads exist, the plant must be vigilant that no excessive new flammable materials be introduced thereby negating the calculations that the cables in those areas will remain operable with encasement. The beat goes on!

Chapter 29

Nuclear Power Plant Operator Training—I Thought You'd Like to Know!

The qualifications for operating a nuclear power plant, including reactor operation are extremely stringent. Basically one would need a high school or maritime academy education, then study for obtaining a state issued fireman's license which would allow one to operate a heating boiler in a commercial building or a boiler in a fossil fuel power plant.

Then comes the fun! Selected candidates, usually from the ranks of the company's production department who operate the fossil fuel boilers are encouraged to try out for the nuclear power plant operators license. Maritime Academy graduates are very good prospects for this program. Navy "Nukes" are a good source of personnel. The nuclear operators license examination program is administered by the Federal Government and given by a branch of the NRC. The training is obtained at the companies training facility prior to sitting for the federal exam. The curriculum encompasses basic mathematics, then escalates to algebra and geometry. The students also study hydraulics, thermodynamics, physics, including nuclear physics, chemistry, including the properties of the Periodic Table of the Elements, electrical properties, both AC and DC, the characteristics and quality of steam generation (Mollier Chart), radiology and the properties of various radiological particles like electrons, neutrons, protons, alpha and beta particles and gamma rays, water quality, heating, ventilating and air conditioning (HVAC), metallurgy, and probably several other studies which presently elude me.

In addition, the students must learn the operation and necessity of up to sixty systems. Piping and Instrument Diagrams (P&ID's) depict the major components, piping, controls, instrumentation, flow process and system logic. Weekly exams are administered and documented for retention. Those students who successfully pass the curriculum now have another hurdle to overcome. They are tested on their knowledge of the plants operating parameters and physical equipment location via "walkthrough" exams with the regulator. Having passed the written and walkthrough exams entitles the students to obtain a federal license which allows them to be employed as licensed nuclear operators at that particular facility they trained on. The license is nontransferable to another nuclear facility. For obtaining their federal license, the operators work three different shifts so as to provide twenty-four hours a day, seven days a week, 365 days a year operator coverage at the facility. In addition they regularly receive refresher training to maintain their skills and also require and receive training on any and all modifications to the facility or to any and all drawings and procedures which have been revised. Nuclear power plant operators probably spend 15% of their physical work attendance receiving training.

All companies who own and operate nuclear power generation capability have, in the past 20 years, installed plant specific simulators equal to or more sophisticated than those at airline facilities. The training classes, the simulator, the curriculum and the constancy of training demanded that the training facility be apart from the power plant.

The simulator is an exact replica of the plants control room, including lighting, tile flooring, color of the control panels, background noise, etc. The advantage of using a simulator is that not only can the students learn to operate plant equipment but any mistakes are not costly to the equipment. People learn by their mistakes. The overriding advantage of the simulator is that accident scenarios can be introduced into the simulator software so the students can observe abnormal responses, up to and including catastrophic equipment failures. The simulator is an invaluable tool in the training of potential candidates and those who already possess operating licenses.

Accident scenarios can be inserted in the software, viewed, and responded to in a stressless environment.

Lest we forget, any modifications to the plants operating systems or equipment requires the exact same modification to the simulators software and in the case of changes in the plants control room panels, the same changes are made to the simulator control panels. Thus we have more documentation and an expansion of configuration control. The simulator would be of little value if it did not replicate the control room at the plant.

All the curriculum the students learned is approximately equivalent to an Associate Degree in Engineering, yet no degree is awarded.

Did I mention that the licensed personnel are required to requalify with the same degree of difficulty as their initial federal testing, on a regular basis in order to retain their license?

Nothing comes easy in the nuclear business.

Chapter 30

The Reactor Recirculation Piping and the Drywell Czar

Tricky title but stick around! All will be explained. First, the reactor has two reactor recirculation system loops, one on each side of the reactor. The recirculation system is one of two ways to increase reactor power output. The recirculation pumps are driven by variable speed motor-generators. Increase the motor-generator output frequency and the recirculation pumps increase in speed, thus increasing the reactor water flow. The recirculation loops are closed systems. They take suction from the reactor annulus (a cavity away from the reactor core) and drive the flow to the lower portion of the reactor (below the core) where the water is forced through jet-pumps (venturis) which induce flow through them. All very technical but easy to understand. More flow means more bubbles swept off the fuel surfaces which generates more steam. (See BWR drawing on next page)

OK, so now you know all there is to know about reactor recirculation flow. So what is this chapter all about? Early in the 1980's General Electric, the nuclear steam supply designer/vendor identified possible problems relative to the integrity of reactor recirculation piping. It seems that this 36 inch diameter piping, manufactured of stainless steel with a specific carbon content in the steel, if subjected to high temperature, radiation streaming and stagnant pockets of radioactive sludge in the piping configuration could develop intergranular stress corrosion cracking (IGSCC). In layman's' terms, General Electric has had reports that piping at several boiling water reactors(BWR's) have displayed minute surface cracking. All BWR's were asked to perform inspections and determine if the cracking appeared at other BWR's. Insulation had to be removed after the plant was shut down and cold. Of importance is the location of the reactor recirculation piping. The recirc. piping communicates with the reactor vessel and therefore is located in the Primary Containment Structure(the inverted light bulb configuration). During normal plant operation the Primary Containment Structure (also called "the drywell") is evacuated of all air and replaced with nitrogen to preclude the support of an electrical fire, should one occur. Therefore, during normal operation, the drywell is not habitable unless one enters with SCUBA gear on.

The QA/QC organization has the responsibility to examine all piping, welds. etc. for acceptance via inservice inspection.

Boiling Water Reactor
Elementary Diagram
Source: U.S. Nuclear Regulatory Commission

Boiling Water Reactor

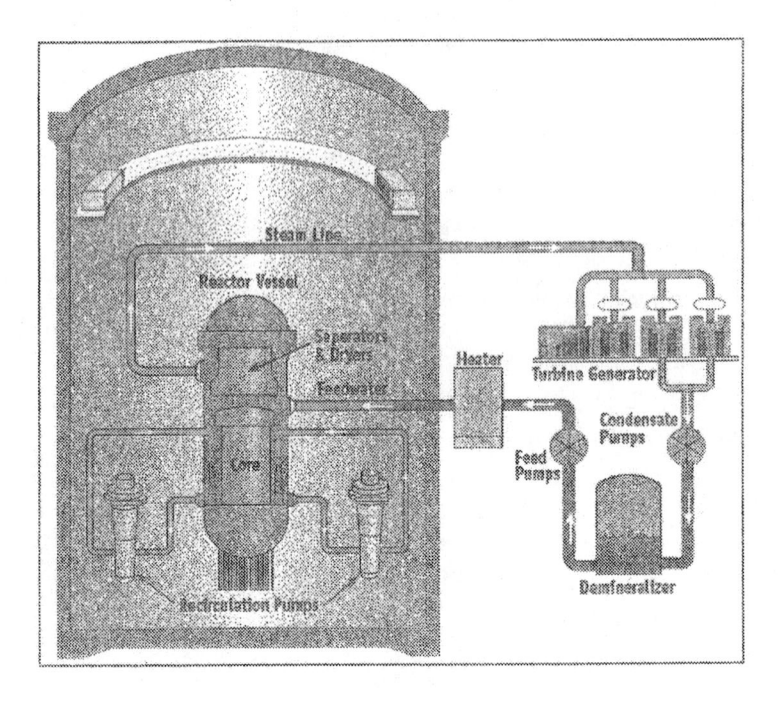

Source: U.S. Nuclear Regulatory Commission

QA/QC identified several indications of IGSCC to the exterior of the recirculation piping and orders were placed to purchase 100% replacement piping on both recirculation loops.

The forthcoming refueling outage would be an appropriate time to replace the piping, if the material were available. Detailed plans and schedules were developed to expedite the piping replacement. Plans were developed to decontaminate the interior surfaces of the piping using a citric liquid solution, driven through the pipes and then separated by osmosis units.

The reasons for decontamination were several, lower radiation exposure to the workers doing the pipe replacement job and also to reduce the radiation levels of the pipe before offsite shipment. The osmosis process was fraught with technical and mechanical problems which were eventually overcome.

So what's with the title, Drywell Czar? It is important to realize that to gain access to the drywell to perform work, there is one equipment hatch and one personnel hatch. The laydown area outside the equipment hatch is relatively small. The recirculation piping, thirty-six inches in diameter and usually between six feet to twelve feet per section in length is difficult to handle and radioactive to boot, upon removal. Personnel working on the removal of the pipe work in Anti-C clothing(coveralls, rubber boots, hoods, rubber gloves, etc.)

Senior management, when reviewing the scope of the work to be performed in the drywell, in addition to the recirc piping, decided work space and laydown area would be limited and critical to schedule success. Additionally, to transport the removed pipe out of the reactor building, once it had been removed through the equipment hatch, a rail track was laid down in the building. In order to coordinate all the work, with various contractors in the drywell, it was determined that a utility manager should have the responsibility for determining work priority and laydown area priority. The Veep of Nuclear Operations, along with the Senior Veep-Nuclear said, "We need a Drywell Czar". His authority would be unquestionable. He would rule the prioritization of drywell work, who will enter, who will work at various elevations, etc. Sub-tier work, i.e. other than that associated with removing and replacing the recirculation piping was given lower priority unless it impacted outage completion schedule.

Significant material had to be removed from the drywell to make room for pipe removal and installation. Huge quantities of contaminated insulation was removed and packaged for shipment and burial. Floor grating was tagged for reinstallation and then removed. Pipe supports and hangers had to be tagged and removed. The laydown and traffic area at the equipment and personnel hatch often looked like Grand Central Station at peak commuter hour. Everybody wanted to do their job now! The master schedule ruled, no ifs, ands, or buts. "Was your job scheduled for today?" "For this shift?" "Is the paper work in order?" "Has the material been staged?" Only yes answers to all the questions were given further consideration. Even then, if the schedule had slippage, so did some sub-tier work slip.

The role of the Drywell Czar was interesting, demanding and all-in-all entirely necessary. General Electric, the contractor hired to replace the recirculation piping, appreciated the fact that they has someone running interference for them to allow them to maintain work continuity effectively.

Of note is the fact that the new pipe joints were welded using remotely controlled state-of-the-art welding machines, strapped to the pipe and operated from outside the drywell structure using monitors.

During the performance of the pipe replacement and refueling the outage manager, along with the Drywell Czar, asked the question "What do we need to do to convince the NRC that the drywell, with all its equipment returned, is ready for closure and startup?" It was determined that the best solution was to develop an equipment checklist which identified all the equipment in the drywell, it's elevation, it's installation, the test procedure used to verify operability, and signoff by the responsible discipline. Only after the entire drywell equipment checklist had been signed off did the company determine it was ready for restart.

The recirculation piping removal and replacement outage, along with the plant refueling was as extremely successful outage, lasting approximately thirteen months in duration. This was deemed successful by all disciplines involved including the regulator and the prime contractor despite the limited work space, limited access to and from the drywell and sparse laydown area from which to stage the work. One of the major indicators pointing to success was the reduced radiation exposure to all personnel working the outage because of preplanning, remote welding and decontamination.

The role of the Drywell Czar, although demanding, was very enjoyable and in hindsight, essential to the success of the job. The author, who was the Drywell Czar, in retrospect saw his performance of that job as rewarding in personal satisfaction and, believe it or not, of minimal stress. There were many plant managers, many refueling outage managers but there was only one Drywell Czar in the history of this nuclear power plant. Long live the Drywell Czar!

Drywell Czar Plaque Photo

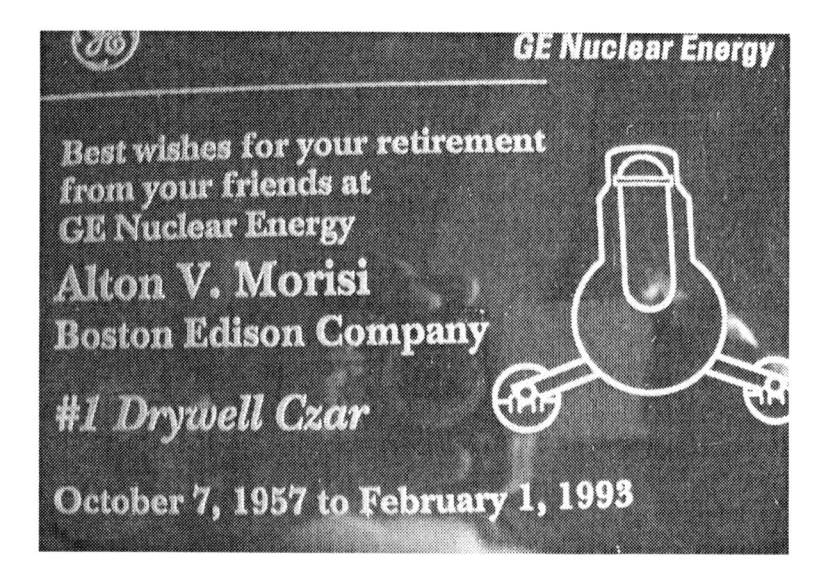

> **GE Nuclear Energy**
>
> Best wishes for your retirement
> from your friends at
> GE Nuclear Energy
>
> *Alton V. Morisi*
> Boston Edison Company
>
> *#1 Drywell Czar*
>
> October 7, 1957 to February 1, 1993

Chapter 31

The Drywell Czar Goes on the Road

Following the successful replacement of the reactor recirculation piping at our nuclear facility, the Sr. V.P. Nuclear called the Drywell Czar with a proposal.

The Senior V.P. Nuclear was in close communication with the President of a sister nuclear utility in the same NRC inspection zone of the country. The sister utility had committed to the replacement of its reactor recirculation piping and was in the mid-range stages of the project preparations and planning. The President had concerns relative to his companies project structure, procedures, manning, materials procurement and particularly with the utilities interface arrangements with the pipe replacement contractors implementation practices.

The discussion between the two executives resulted in my utilities Sr. V.P. Nuclear offering the services of the Drywell Czar to the sister utility as a "have been there, done that" expertise for oversight of the planning and proposed implementation to date.

How could one refuse such an offer? The Drywell Czar was a long time industry buddy to the President of the sister utility, having come up through the ranks of nuclear power positions concurrently and maintaining a utility contact through the years.

The Drywell Czar packed his bags in his company car and headed for another challenge. The first problem was that the sister utility executive had not informed his management cadre that a Drywell Czar was coming to look over their shoulders and report his findings to the president directly. That was just what they needed, a "dime dropper" in their midst or a "rat in the woodpile". I was greeted with open arms by the president who proceeded to call his team players to his office and introduced me to them. He tried to downplay the "spy" bit by iterating that I was there to observe their organization, its composure, its procedures, its planning and tentative schedule of planning, purchasing, staging, implementation and acceptance testing of the project. I was cadred with the Sr. Maintenance Engineer and the Planning Group Leader for what was supposed to be cohesion of thought.

Problems arose from day one when the Vice President of Operations at the facility took exception to my presence. I had many discussions with this individual who had overall responsibility for the implementation of the project. I finally convinced him that my being there and my findings were from the standpoint of assisting their efforts because of my life experiences. I would inform him of any concerns before I reported them to the President. That was the clincher. He and I then had an excellent repoire and we accomplished much to improve the projects performance. My "in" was confirmed when I was asked to join the after work nine-hole golf group.

Their project relied heavily on two aspects: (1) that the constructor was allowed carte blanc authority to perform per his construction program and (2) the constructor would interface with the utility by embracing the utilities everyday work practices and procedures.

Both of these aspects are noteworthy on face value. In essence the project was to be implemented by the contractor being given considerable authority. In plain English this said "Here is what I will do, now get out of my way and trust me!" Oh boy! The road to hell is paved with good intentions! The main problem is this. The utility, at all times, is responsible for all aspects of the operation of its nuclear facility, both to the NRC and to the state Public Service Commission (PSC) where the plant is constructed. In this case the NRC/utility interface is of minimal significance as long as all applicable procedures are followed explicitly. But the most significant interface is with the PSC. As always, once an outage has been completed, the utility pulls together its plans and schedules, progress reports and performance records and goes before the PSC to explain its performance and demonstrate the cost effective and prudent implementation of the outage. The PSC is responsible for overseeing that utility customers receive fair value for their energy dollars. The utility must prove due diligence so that the PSC will grant monetary reimbursement for all outage expenses.

Back to the project planning!

A utility hires a competent constructor to implement the physical work but the utility must have final say in performance implementation. This is not a "turnkey operation". Call me when you are done! The constructor is not held responsible for outage slippage, by the PSC, the utility is.

My overview concerns were that the utility was delegating and relying almost exclusively on the constructors "trust me" performance. I informed the VP-Nuclear Operations and the President of Nuclear of my concerns.

The VP Nuclear Operations was reluctant to go along with my findings. His philosophy was that "we hired the contractor to do the pipe job so let's get out of his way and let him do the job". The President was open to my concerns. He could envision his presence before the PSC testifying under oath that the outage was performed prudently by the contractor under the watchful eye of the utility.

Another finding and concern by the Drywell Czar was that the utility employees, outside of the planning and scheduling group, were not gung-ho!

They were sitting back watching the construction contractors personnel doing their thing and relying on the "trust me, I'll get it done on time" philosophy.

I recalled that at our utility, the utility personnel who had responsibility to the project were all over all aspects of the project with the contractors personnel. Another observed finding was that the contractor was conducting the daily update/problems briefing meetings, and not the utility. From our point of view the utility conducts all briefing meetings daily with major input from the contractor personnel. Our philosophy is, we are the captain of this ship, it is our license and you, the contractor, are a player in this project.

Several other findings by the Drywell Czar were noted to the sister utility for consideration and possible implementation. Most consisted of work practices, physical space to integrate and perform the work, procedure requirements and staffing of the project by the utility.

The Drywell Czar report was submitted to the sister utility President from the Drywell Czars Sr. VP-Nuclear after having discussed the findings in briefing meeting with both utilities executives.

The Drywell Czar returned several times to the sister utility during the implementation of the project. This interface was not an on-the-premises at all times affair.

The President of the sister utility took the findings and implemented a majority of them, where it was practical without reinventing the entire work process as it was normally performed. For the most part the project was of moderate success. The contractor, as all contractors are prone to do, faltered on several aspects of the plans and schedules. This was not surprising. Most projects have blips, unexpected problems, etc. The fact is that the sister utility became more intricately involved in the contractors performance, the contractors problem solving abilities and in essence did what it was licensed to do, manage a nuclear facility at all times and all stages of operation.

The bottom line of this whole Drywell Czar on-the-road experience assisted the sister utility perform its reactor recirculation pipe replacement project with insight from past experience, which often helps immensely.

Of note is that the President of the sister utility resigned shortly after the completion of the project but before the PSC hearings on prudency were instituted. There is no inference here that his management oversight was remiss. Sometimes work stress becomes overbearing. The entire subtleties of this book refer frequently to human burnout. Sometimes the job is just too overbearing and people just chuck it in! We, the former Drywell Czar and the ex-president of the sister utility have crossed paths later in our careers, I found him to be content and revitalized in his new profession, within the bounds of nuclear power but out of utility management.

Long live the ex-nuclear power jocks!

Chapter 32

What Happens if Your Reactor Vessel Springs a Leak?

Mention was made several times that a shutdown reactor is in it's safest mode, per definition. In the chapter on TMI, the reactor core was allowed to become partially uncovered by the loss of water inventory in the reactor vessel and the core overheated, melted, and slumped into the lower portion of the core. Also mentioned was the role of water in the reactor, it cools the fuel, it slows down streaming electrons so they can be captured by surrounding fuel to generate more heat and more steam and it acts as a shield against radiation exposure.

The reactor recirculation system piping configuration states that it communicates with the reactor vessel as a closed loop. It takes suction from the reactor and drives the water up through the core to increase power.

Wherever water is used, in the home, in industry, there is always concerns about leaks. Water tends to leak to a lower gravitational level whenever it can. The same goes for water in a nuclear power plant. Plant systems which process water or any liquid are susceptible to leaks.

So what happens if we spring a leak somewhere in the reactor vessel connecting piping? The plant is designed to overcome any conceivable leak including (now get this lingo) a guillotine break in the largest pipe which connects to the reactor vessel. That pipe is the reactor recirculation piping in the drywell. Should something cause this pipe to sheer off completely, exposing an open thirty-six inch pipe leakage, the plant Emergency Core Cooling Systems (ECCS) are designed and prepared to flood the reactor vessel to maintain reactor core coverage and prevent the overheating and destruction of a portion of the fuel core.

"What are these systems?"; "How do they work?"; "How many are there?"; "Do they need electrical power to operate?"; "If yes, what happens if a leak occurs during a storm which knocks out electrical power to the plant?"

To get semi-technical, the term used to describe the leak is called a Loss-of-Coolant-Accident (LOCA). Now for some answers to those pesky questions.

There are several systems which directly or indirectly are designed to shut the reactor down, release built-up steam pressure, poison the core if necessary to shut the reactor power off and to flood the reactor vessel to maintain water inventory above the top of the active core.

Let's discuss the leak first. The reactor recirculation piping has isolation valves primarily to block the flow of water in the recirculation system. Also, the recirculation pump suction comes from an annulus, a cavity within the reactor vessel about two-thirds of the way up the core. One could describe the annulus as a spillover section in the reactor vessel, so if the recirc piping sheered off, the reduction in reactor vessel water inventory would only drop down to the top of the annulus and one would have to provide replacement water to raise the level to the top of the fuel core (about one-third of the total fuel core height).

That said, let's look at ways that are designed to reflood the core. First there is the High Pressure Coolant Injection System (HPCI). It is so named because it is the first system to operate to reflood the vessel. The HPCI system consists of a steam driven turbine which takes suction from a huge tank called the Condensate Storage Tank. The reactor has been scrammed automatically upon detecting low water level and the steam, still being generated and now not going to the turbine generator to make electricity, is directed to the HPCI steam turbine to drive a pump which transfers water from the condensate storage tank to the reactor vessel to reflood the core. This continues until the steam generated in the reactor can no longer support the steam needs of the HPCI turbine.

The next system to step up is the Reactor Core Isolation Cooling system (RCIC). This also has a steam driven turbine to pump water. However the RCIC is a lower flow capability system which can operate on low pressure steam with high moisture content in the steam. This is the second core flooding system for recovering water inventory. It too takes suction from the condensate storage tank.

Another system is called the Core Spray System and it consists of multiple electrically driven AC pumps which take water suction from the condensate storage tank and sprays water over the top of the fuel core.

What happens when we use up all the water in the condensate storage tank? Another tricky question! Remember, the drywell is a structure in the shape of an inverted light bulb with reinforced concrete backing. Now picture this. The inverted light bulb has several legs(large pipes) which go from the bottom of the drywell down to a Torus(a big bagel shaped circular tunnel) which sits below the drywell, is large enough to drive a car around inside, and contains water inventory. The Torus is also known as the wetwell. So we have a drywell, a building, and a wetwell, an enormous circular tunnel of steel construction. The Core Spray system and the RCIC system can transfer their suction needs from the condensate storage tank to the torus for water supply. Water that leaks into the drywell flows down into the torus (wetwell).

"What happens when all the hot water flows down into the wetwell?"

There are multiple AC electric driven pumps which are designated as the Residual Heat Removal System (RHR). The function of these pumps it to take suction from the torus wetwell and pass it through heat exchangers to cool the water and discharge it back into the reactor through the recirculation system piping (into the recirculation piping which does not have the sheer break).

These systems, in combination or individually, can maintain reactor water level above the active fuel core and also cool the reactor water. Did I mention that these systems operate, via instrumentation logic, automatically?

All emergency core cooling systems are surveillance tested on a regular basis, with the exception of final injection of water into the reactor. Those final tests are performed when the reactor is shut down for refueling.

Now that covers most of those nagging questions we started out with, but we must consider the reliability of the electrical power to drive the Core Spray and RHR pumps.

Should the ultimate event occur, we have sustained a large pipe break and simultaneously suffered a loss of power. The technical definition for this condition is "Loss of coolant accident" (LOCA) with "Loss of offsite power"(LOOP). We have a LOCA with a LOOP. No problem. The plant has not one but two Emergency Diesel Generators. One generator can supply the power, in case of an emergency, to load and drive the Emergency Core Cooling Pumps and shut the plant down. The second emergency diesel generator is for redundancy and power reliability. Emergency banks of batteries are available to provide power to instrumentation for shutting the plant down and operating the Emergency Core Cooling System logic.

If the control rods in the reactor fail to completely shut the reactor down, the plant can inject sodium pentaborate (twenty mule team stuff) into the bottom of the core with the Standby Liquid Injection System to poison the core and shut down the reactivity. Should reactor pressure escalate, the Automatic Depressurization System can be operated to relieve steam pressure by opening relief valves which are backed up by safety valves.

As a final resort, there are connections which will take suction from the ultimate heat sink, the ocean, to supply water inventory to the reactor vessel. When that happens, we are injecting briny salt water into the stainless steel system piping and the reactor vessel. After that, we can padlock the plant and go home for keeps.

Chapter 33

The Nuclear Power Game, the Author and Burnout

The outline for part of this book resided in an expandable binder for the better part of fifteen years or more. The material is dated relative to the present requirements for major modifications to backfit the remaining operating nuclear power plants. The nuclear power industry in the US has had and continues to have an admirable record for availability, power generation and especially, safety to the general public for greater than forty-five years. Little is ever read or reported on national news broadcasts to cast a negative light on the industry. As a matter of fact, the energy bill presently stalled in Congress contains provisions for assisting the nuclear industry to revitalize the construction of new nuclear generating facilities so as to become less dependent on foreign energy. The following article appeared in several major newspapers in April, 2004.

"WASHINGTON—Seven companies have agreed to jointly apply for a license to build a new commercial nuclear power plant, the first new reactor application to be filed in three decades, the companies announced Wednesday.

The five energy companies and the two reactor vendors emphasized that none of the companies have made a commitment to build a new plant, but are taking the move to test the government's streamlined licensing process.

The companies intend to commit $7 million a year to the effort under a cost-sharing program with the Energy Department. The goal is to get license approval from the Nuclear Regulatory Commission by 2010."

New friends occasionally ask me what I did for work. When I tell them I was employed for greater than thirty-five years in power generation at a public utility and twenty-five of those years were in all aspects of nuclear power plant performance, they roll their eyes. I'm not sure they are surprised, disappointed or non-committal. Presently retired and still residing within the influence of another nuclear power plant, I seldom see or hear disparaging remarks either aired or in print. It appears that the ugly giants have become good citizens in their neighborhoods. Some of my fellow workers occasionally get together to commiserate about our employment days, curse the company for the misery it caused us and our families and turn around and toast the company for our rather satisfactory lifestyle in retirement.

I guess we relish the good days and detest the bad ones.

Discussions often strays to the fact that we were pioneers in a new technology, nuclear power generation, similar to the astronauts in space technology. We too literally flew by the seats of our pants. Retrospect would support the facts that as abnormal events occurred at our sister utilities, the NRC had to institute demands for improvements in facility safety as mandated by their responsibility to protect the general public from nuclear power harm.

Several chapters reiterated the negativity of fossil fuel executives and their failure to comprehend the nuclear game requirements but persisted in imposing their antiquated management styles. So what else is new?

There were many days of glory and also many days of gloom. The nuclear power industry should have continued to grow and unburden the need for excessive amounts of foreign oil but two incidents altered the history of nuclear power. Presently some utilities are processing requests for extended operating licenses beyond the original forty years of plant operation. Others are evaluating whether to do so or shut down and decommission their plants. Either way, nuclear power is stagnant and declining. There have been no orders for new nuclear power plants post TMI to the present time.

The resultant expenditure of millions of dollars for modifications plus the increased personnel staffing demands to meet rising standards of excellence have literally priced nuclear power into a noncompetitive posture relative to fossil fuel operating costs. What was once an initial expenditure of capital for construction and then competitive operating costs went south for nuclear power, thus negating its advantage over fossil fuel performance.

Hopefully there will come a day when nuclear can become competitive and a viable source of electrical power production. The main theme, as expressed vehemently many times is the inordinate demands unilaterally imposed by the regulator upon the power companies to expeditiously implement numerous major backfit modifications to assuage the general public that nuclear power is safe. Nuclear power was safe, nuclear power is still safe. The mishaps at TMI, Browns Ferry and Chernobyl were indeed serious incidents but to repeat, no one was killed, no one was overexposed in the U.S. It is the authors contention that most of the backfits should have been implemented, but the need for expediency was overkill. Too much was demanded across the engineering spectrum in order to pacify the regulators. Safety wasn't compromised but it might well have been under such stressful working conditions.

Back to the title of this chapter. The author, a graduate engineer, worked in the electrical systems department, both in the field and was at a fossil fuel power plant when the company signed a purchase contract in the mid 1960's to have a nuclear power generating station built. Senior managers were asked to submit recommendations of in-house engineers who should be interviewed and considered for cadre staffing. I was on the list and interviewed for a mid-level managers position as the Instrumentation, Control and Power Division Manager. The fact that the plant was to be constructed within easy commuting distance from home also helped make the decision to accept the companies offer. The cadre of in-house personnel were on the road training for this new technology for the better part of two years in various part of the country. Upon their return, they next oversaw the construction of and provided input to the design of many plant systems based upon past experience. Life was great! The challenges were new and exciting. Everyday brought new experiences that required attention. The job was a pleasure.

Technicians were hired to complete the plant staffing, where required. Laboratories were designed and equipped and test procedures for calibration and surveillance testing were written, field tested and approved. Systems were turned over to the client for testing and acceptance. The regulators were novices just as we were and we all learned our new jobs together.

The power plant was completed, granted a license to start up and then issued a full power license. The "new baby" started to make money for the corporation. Everybody was literally on cloud nine. We did it! Everybody looked forward to the next two years of operation and the next new adventure, the first refueling outage. Somewhere along about three years after startup, the regulators demeanor started to stiffen. Configuration control reared it's ugly head and a few deficiencies were identified. At that time the engineering support organization consisted of a few individuals in a non-formal structure without detailed work procedures.

In 1975, the company, having contracted for a second nuclear power generating station on the same site, incorporated the on-site ad hoc engineering group with a newly formed engineering department whose primary responsibility was to review the design of plant #2. Several equipment problems at the original facility resulted in some modifications and backfits over the next couple of years.

Then along comes Browns Ferry followed by TMI and the roof caved in. The Engineering Department increased staffing in all disciplines several times over. Reorganization resulted in the separation of Nuclear Engineering and Nuclear Operations into separate departments in the nuclear organization.

Previous chapters highlighted numerous problems with the organizational structure from the executive level on down. The resident NRC Inspectors were assigned to all the nuclear facilities and those with less than exemplary performance were gifted with two or more resident inspectors.

The author was promoted to Manager of Nuclear Operations Support and as mentioned, all responsibilities without a dedicated home were incorporated into the new department. The span of control spread out to at least seventeen individuals reporting directly to the manager. Impossible! They say if you can't stand the heat get out of the kitchen. So I transferred back to the Operations Department as Assistant Refueling Outage Manager. What appeared to be a smart move at the time later on became another untenable position, that of Refueling Outage Manager for about 35% of the 33 month "show cause order" shutdown and refueling. Fortunately, I was able to eventually vacate that position before the last 10% so I returned to a comfort zone.

After TMI and up to and including the Refueling Outage Manager position, the jobs were literally years of rising stress and frustration. There were absences from the children's birthdays, wedding anniversaries and children's sicknesses. The pager went off incessantly, the telephone rang at all hours of the evening and nighttime. I was literally married to the company facility as were many others. Senior plant management changed like the seasons and probably as often. The regulator became disheartened with plant management performance and everyone, including those who had toiled in the vineyard since sunup, were painted with a wide brush.

After startup from the "show cause order shutdown" things were not the same. Senior plant managers bailed out, having rescued the plant from further difficulties. New people were brought in who displayed elevated vigor, not having suffered the previous stressful years.

As if a godsend, the company announced they were offering early retirement to those employees who met years of service and age requirements. I was short by about 1 year from receiving full benefits. Quietly, I transferred over to staff assistant to the Veep of Operations who himself was newly hired. I developed the processes for managing the Long Term Plan and got it off the ground.

A new project mandated by the NRC required all nuclear plants to evaluate their facilities and document where the organization should spend it's limited maintenance budget for the most "bang for the buck" along with nuclear safety. I was a natural for this project as were several other employees waiting for the opportunity to retire. I joined the project and remained there in a non-stressful position until I acquired the necessary numbers to accept early retirement. Low and behold, upon my early retirement, I was asked to return, as a contractor, to complete the project evaluations for maintenance reliability. That project lasted just months short of two years, whereupon I was then eligible for early social security benefits.

I accepted my lump sum retirement package, my severance benefits, my contractors wages, my social security benefits plus many weeks of unemployment benefits and moved south to enjoy year-round golf.

Would I do it all over again? I'm not sure. Probably not. There were other satisfying jobs in the company, prior to venturing into the nuclear game, which I could have performed with minimal stress and negligible burnout.

Discussions with friends who still toil in the nuclear power field reveal that stress levels are down somewhat because modifications are at a minimum but replacements/upgrades of equipment are on the increase. Senior management still assesses operating costs versus staffing and sometimes shoots itself in the foot (still) by reducing staffing and then trying to climb the mountain with fewer sherpas. Chitchats with a relative who is employed at a nuclear facility reminds me that it's deja vu all over again, same problems, same management edicts, same personnel turnover because of the inability to see the trees for the forest. No one ever looks at history but tries to operate as if it was day one! Some things never change!

Chapter 34

Nuclear Power in the Generation Mix in the Future

There seems to be a stir relative to the future of nuclear power plant construction. Several articles appear in local newspapers and their frequency give hope that the construction, licensing and operation of nuclear power in the USA may again come to pass. The NRC is actively involved along with some nuclear power plant designers to initiate the licensing process for future construction. The time span, hence forth, is probably somewhere along about ten years before a new plant is started up. Consideration for standardized design of two or more models is afoot. General Electric and other designers are deeply involved in the process of standardization to facilitate an easier approach to getting licensed by the NRC.

Just recently, Progress Energy of Raleigh, North Carolina, announced plans to construct a new nuclear plant at their Shearon Harris site and are also seeking a location for a new, possibly nuclear, plant in northwestern Florida. Hope is on the horizon!

Maybe the rise and fall of the nuclear power industry will, like the phoenix bird, rise out of the ashes and reappear. Maybe the title of this book should be: *Days of Glory, Days of Gloom and Back Again*, along with *The Rise and Fall and Rise Again of the Nuclear Power Industry*.

The US needs to address the use of alternative fuels for power generation. The economy moves and grows as a result of electrical power capacity. Nothing drives progress more than the ample supply of electrical power, not only in this country but throughout the world. Third world nation countries will only rise above their present social status by developing more electric power capability. We in the USA need to reduce not only the dependence on foreign oil but also the consumption of oil for our electric power needs.

Let's hope that future generations find not only nuclear but alternate means for the generation of electric power and for the transportation needs of our population. Nuclear can go along way in helping alleviate one of these conditions.

Printed in the United States
57026LVS00002B/251